D0463446

The Face

Also by Daniel McNeill

Fuzzy Logic (with Paul Freiberger)

The Face

Daniel McNeill

Little, Brown and Company
Boston   New York   Toronto   London

First Edition

The author is grateful for permission to include the following previously copyrighted material:

Excerpt from "In a Station of the Metro," by Ezra Pound, reprinted by permission of New Directions Publishing Co. Copyright © 1956, 1957 by Ezra Pound.

Photograph of *The Kiss* by Constantin Brancusi copyright © Graydon Wood, 1994, The Philadelphia Museum of Art. Courtesy The Philadelphia Museum of Art: The Louise and Walter Arensberg Collection, 1950; photograph of *Castiglione* by Raphael copyright © R.M.N.; photograph of *Mona Lisa* by Leonardo da Vinci copyright © R.M.N.; photograph of *Sarah Siddons as the Tragic Muse* by Joshua Reynolds courtesy of The Huntington Library, Art Collections, and Botanical Gardens, San Marino, California.

Drawings on pages 6, 7, 31, 64, 68, 88, 89, 179, 298, 299 by Sharon Constant, Visible Ink, Oakland, California.

Photographs on pages 183, 184, 185, 186, 187, 188, 189 copyright © Paul Ekman, 1990.

Library of Congress Cataloging-in-Publication Data
McNeill, Dan.
      The face / Daniel McNeill. — 1st ed.
         p.   cm.
      Includes bibliographical references and index.
      ISBN 0-316-58803-2
      1. Face. 2. Face perception. 3. Facial expression. I. Title
      QM535.M27   1998
      611'.92 — dc21   98-15147

10   9   8   7   6   5   4   3   2   1

MV-NY

Designed by Barbara Werden

Published simultaneously in Canada by Little, Brown & Company (Canada) Limited

Printed in the United States of America

To the memory of my parents,
Harry and Maureen

Contents

Everything is in the face.

CICERO

Prologue

I WAS walking down a street called The Navel of the World, one of two paved lanes in town. Banana fronds hung prostrate in the mugginess, shrouding homes in wanton green. I felt laminated. A man on horseback clipclopped past, singing, a child on his lap. At the harbor I gazed out at the rowboats and the vast ocean. I was on Easter Island, deep in the South Pacific, two thousand miles from anywhere.

I hiked out of town on the red-dirt road, past the shed-like disco and around a cemetery. Wild horses grazed by the surf and grassy slopes swung up to a volcano. Then I spotted them in the distance: tiny upright blurs. They stood on rock platforms by the blazing sea, staring in at meadows where clans once gathered.

The great stone busts of Easter Island are among the best-known faces on earth, and they're a potent sight. Each preserves the features of a man lost to history. And the mind too, for I peered into these basalt heads and sensed thought. The idols smolder with dominion, and actually frightened nineteenth-century traveler Pierre Loti. They are lords disdainful of all, even the rain-drops which have almost erased them.

Back in town, as I idled through the open-air market where women and children sell carvings and fruit, the faces of Islanders were everywhere. They grinned, grew solemn and hopeful, lit up, assured, thanked. Their eyes queried, danced, and glowed, and these faces were more compelling than the idols'.

But of course. The living face is the most important and mysterious surface we deal with. It is the center of our flesh. We eat, drink, breathe, and talk with it, and it houses four of the five classic senses. It is also a showcase of the self, instantly displaying our age, sex, and race, our health and mood. It marks us as indi-viduals. It can send messages too elusive for science, so far, and it bewitches us with its beauty. The Trobriand Islanders deemed the face sacred, and well they might, for it is our social identity, compass, and lure, our social universe.

Our minds light on the face like butterflies on a flower, for it gives us a priceless flow of information. When Robert Louis Stevenson crossed the Atlantic in steerage, he sought recognition as a gentleman and marveled sadly that fellow passengers looked at his face and not his uncalloused hands. We even watch the faces of ventriloquists' dummies, though we know they're made of wood.

Indeed, faces grip us from birth. Babies just nine minutes old, who have never seen a human countenance, prefer a face pattern to a blank or scrambled one. Similarly, monkeys raised in utter isolation can identify photos of their own kind. "You come into the world knowing what a face looks like," says psycholo-gist Vicki Bruce, perhaps the world's leading authority on face recognition.

Faces wisp up out of chaos for us, as if beckoned. We claim to spot a man in the moon. Scrooge saw Marley's face in a door

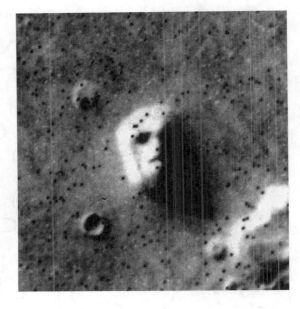

The Face on Mars.
Courtesy National
Space Science Data
Center.

knocker. The three-prong wall socket suggests a wide-eyed shriek, and the pious conjure the Virgin Mary from Eucharistic hosts and photographs of the sky. In Lorrie Moore's novel *Who Will Run the Frog Hospital?* Sils detects the faces of Napoleon Solo and Ilya Kuryakin in her toenails. The "Face on Mars" has become a tabloid favorite, and one excited 1996 article claimed a lost civilization sculpted it around 2000 B.C.[*]

Indeed, we can draw almost any closed figure — even just a wobbly circle — and make it a face by daubing an eye at the right place. We are so attuned to faces we'll build them from a squiggle and a dot.

Yet what happens when we look at a real face? Intriguingly, we seem to gaze through it. It half-vanishes, like the pages

---

[*]NASA has hidden this news from the public, presumably to quell ardor for the space program.

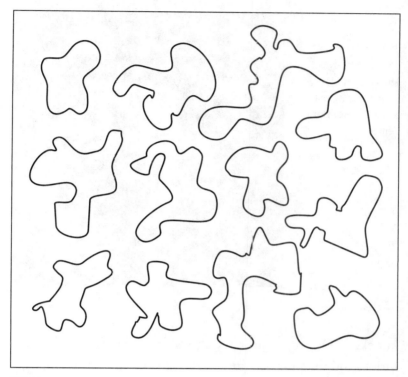

Wriggles.

of a book, so we view soul more than contour. Hence we often treat the face *as* the self. When Charles Darwin's ten-year-old daughter Annie died in 1851, he mourned the loss of "her dear joyous face." Marguerite Yourcenar, in a love crisis, wrote, "Absent, your face expands so it fills the universe." We crave the faces of those we care for and keep their pictures on desks or walls, drawing spirit from the selves within.

The brain has special sections for the face. They automate our reading of it, whispering their conclusions into our conscious minds — he's upset, she's beautiful, that's Steve. We see the face without quite seeing it, and its mysteries glide beneath us.

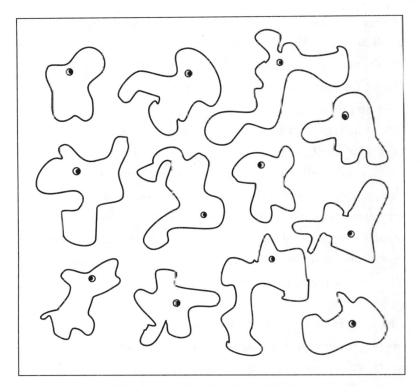

Wriggles with Eyes. (After Scott McCloud, *Understanding Comics*)

   This effect both backlights and bedevils the four great realms of the face: its anatomy, singularity, expression, and beauty.

   Anatomically, the face is a land of veils. Its shape speaks of the primordial sea as well as savanna technology, and its details hold nectarous secrets. What makes our ear flaps convoluted? Why do we have lips, and what good are eyebrows and hair? These features perform deep and splendid services, as we'll see, but others like the chin and nose are more puzzling.

   Every face is unique. Six billion adorn the earth, and the number of possible faces may exceed the number of particles in the universe. The face is a christening in flesh, and people who change their faces with plastic surgery often feel they've lost part of themselves. We excel at recognizing faces, and can tell a man's from a woman's at a glance, without knowing how. Yet almost no one can accurately describe a face — a curse to novelists and detectives.

The face is an uncanny semaphore. In life and in some fictions like *Jane Eyre*, it issues messages of startling depth and infinite hue. We rely on these signals constantly and willy-nilly, for almost none of us can define them. We are reading a language we cannot articulate and may not consciously notice. Yet we regularly feign these cues. Deceit pervades animal communication, and even chimps can lie with their faces. The face is both ultimate truth and fata morgana.

Beauty has touched everyone from Socrates to thugs in Chandler mysteries. Ovid called it "a gift from the gods," and people throughout the globe seek its siren power for themselves. It has always been a breathtaking enigma, and its dazzle has left many artists joyously face-struck. Science has shown it to be a stranger concoction than most people realize, and researchers are still probing why it matters, and even what it is.

Many mysteries of the face don't lie in the face at all. They lie in the mind, in the ways we respond to the face, represent it, hide it, adorn it, and try to remake it. The face in the mirror has its secrets, as does the tattooed face and the actor's face, the tyrant's icon face and the face of God.

It is a magnificent surface, and in the last twenty years we've learned more about it than in the previous twenty millennia. We have commenced to map the world of the face, and it has more surprises than anyone would have guessed.

1 Genesis

The apparition of these faces in the crowd;

Petals, on a wet, black bough.

EZRA POUND

One   A Tour of Parts Unknown

IN his *Travels* (1356), Sir John Mandeville found the Andaman Islands rife with sensational beings. Headless humans strolled about with eyes and mouths on their chests. He saw noseless, sheet-faced citizens with punchhole eyes and lipless mouths. On one island, residents had a huge upper lip that shaded their faces as they drowsed in the hot afternoon. Elsewhere, tongueless dwarves with hard, grommet mouths sucked up food through a straw, and people walked about with ears that hung to their knees.

One scholar said Mandeville never traveled further than the local library, and it's almost certainly true. Faces like these inhabit the same realm as centaurs and flying monkeys, and for the same reason. They don't work.

The real human face is a glory of function, yet strange in ways beyond Mandeville's imagination. Indeed, it sets us apart from even the Neanderthals. For instance, it is flat, an extraordinary fact in the snarling animal world. Our mouths, noses, foreheads, and chins are almost unique. Males sprout beards and mustaches, unusual among primates. Our hair grows so long we regularly cut it, unlike any other creature.

It is a singular structure and it teems with subtleties. Indeed, it resembles a mansion full of invisible servants, little Ariels like the eyebrow performing tasks we never realize. It's also a zone of sensuality, of surging lips and gossamer hair, dancing eyelashes and pupils bright with sin. Such cues can be obvious, but they can also work sub rosa, spinning delight out of apparent air.

Of all items we see in daily life, the face most urgently needs a tour, for it is an enchanted terrain, one that both engages us and sedates the curiosity. From the eyes down to nose and mouth, and out to the frame of ears and hair, it is a playground of secrets. Some are shallow and some very deep, and the most basic of them goes back to the early days of animal life itself.

The Early and the Odd

Why have a face?

We don't strictly need one. Many creatures, like sea urchins, starfish, clams, jellyfish, and protozoa, disdain it entirely. Others have partial faces. The microscopic rotifer has a pair of eye-spots on a rod in a feeding cup, an almost faceless face. The face of the sea anemone is all mouth, and of the octopus, two peering eyes. Snails have tiny mouths and eyes on stalks that wave over their heads.

Yet faces are amazingly common in the animal kingdom. Jaguars, salamanders, and hawks have them, as do all insects, fish, reptiles, birds, and mammals. Why them and not the jellyfish?

The answer lies in evolution, the treasure chest of meaning for anatomy. The faces of everyone — Marilyn Monroe, Ronald Reagan, William Shakespeare, Cher — began in the sea.

A true face bundles mouth and sense organs, and it may be older than shell or bone. Geneticists say multicelled life arose around 1.2 billion years ago, but hard fossils don't appear until much later, around 544 million years ago. Soft-bodied creatures like worms and the weird, feather-like Ediacarans scuttled about in this vast eon, but their remains are scarce. The first face likely coalesced toward the end of this period.

When shelled creatures suddenly did arrive, they were spectacular. It was the great Cambrian carnival of life, the zoo-logical equivalent of the Paris art world in, say, 1910, and mem-orable faces adorned the seas. The tiny *Opabinia*, for instance, boasted a tentacle like an elephant's trunk and five mushroom-like eyes. It is so weird that, when first shown to an audience of paleontologists in 1972, they burst out in laughter.

But other creatures were more recognizable, like the slith-ery worms with eyes and mouth in front, and they show the origin of the face. It is the child of motion. When an animal swims regularly in one direction, the head becomes its leading edge. A forward mouth swallows food easily, through simple momentum; a mouth astern would recede from it. The head also contacts nonstop novelty, so the sense organs cluster up front, like the guidance system of a missile. There they reveal the future, an animal's fate in the space ahead.

Vertebrates like fish have four times more basic structure-mapping genes than invertebrates, and more intricate heads and bodies. And with fish, significant brains become common. The brain audits the facial senses. In theory it could go anywhere, like the central chip of a computer, but biological wiring is frail. So it too lodged up front, like a pilot before an instrument panel. And since the mass of neurons makes a choice killing point, fish evolved a hard covering for the brain, a skull or head-shield. Bone became the foundation of the face.

If you could take the stars in the Orion and shift them about at whim, the constellation would quickly vanish. Orion is more than Rigel, Betelgeuse, and its other stars. It is a pattern, and so is the face. In fact, we see the same array of mouth, nostrils, and eyes in creatures from eels to Einstein. It has been a marvel of hardiness, outlasting mountain ranges. That means evolution has crushed other designs.

What's made it so durable? Why, for instance, does the mouth always lie below the nostrils and eyes? And why don't we have eyes in the back of our heads, so we could see the whole world at once?

The face has a master sculptor: the quest for food. Hence the mouth dominates everywhere, in toads and foxes, caymans and wildebeest. It is the portal where an animal assimilates the world, begins to change it from nonself to self. Hazards abound here and care is paramount. So three checkpoint senses — taste, smell, and sight — lurk nearby to reject poisons and generally tell ambrosia from ash. The taste buds lie within the mouth, the nostrils sit just above, and the eyes perch a tier higher.

Why are the latter two above the mouth? As it happens, this placement yields many rewards. For vertebrates in general, it sets the eyes above falling food and out of the body's shadow. Fish especially need eyes oriented to sunlight, which fades even a few hundred feet down. The arrangement also lets land animals gobble morsels from the earth, sniff rising aromas, and view snout and ground at once, instead of snout and sky.

Some creatures boast extra senses on the face. Blind cavefish have ridges of tiny rods which detect the ripples in water from moving prey. The dimples of pit vipers like rattlesnakes can register shifts in heat as slight as 0.002° F. and help them strike rodents in dark burrows. And sharks have a bulb of flesh jutting out between their eyes and low-slung mouths. Long a puzzle to zoologists, this nosecone contains the ampullae of Lorenzini, which sense electrical pulses from living creatures. To a shark, all food is like a parolee with a radar bracelet.

Strong forces have fixed the face pattern, yet species can stretch it. Carnivores like cats and bears have frontal eyes like headlights to give them binocular vision, which sharpens their sense of 3D, makes them better hunters. But this pairing narrows the visual field, so they often compensate with swivel necks—an owl's can turn some 270 degrees—and eyes that rotate in their sockets. Like all higher primates, humans hew to the carnivore model and have necks and matched, movable eyes. We don't have eyes in the back of our heads because by turning our necks and eyes we can see the whole panorama anyway.

Prey, however, need a faster early-warning system. Many, like gazelles, have eyes on the sides of their heads, to scout a wider range and spot cheetahs earlier. Most fish employ this sentry strategy, and in fact their lenses can bulge through their pupils, giving them a near 360-degree view. Some bottom-feeders take a more drastic tack. Their faces split. The ray, for instance, has a mouth and two nostrils on the smooth, broad belly where it eats. But its eyes lie on top. Its enemies can only come from above, so eyes below would be a gift to barracudas.

Nature smiles on other deviations. Nostrils atop the skull seem pure Mandeville, but whales and dolphins have blowholes there to breathe more easily when they surface. A long, liana nose is bizarre, yet the elephant grew one because its head can't reach the ground. The great flanged face of the hammerhead shark is one of the oddest in the vertebrate world. It may have arisen to spread the nostrils farther apart, increasing the difference between odor levels in each, so these sharks could better track the origin of delectable scents.

Does any animal have eyes on the back of its head? The shrimp *Rimicaris exoculata*, which lives near seafloor vents in the Pacific, comes close. It has eyes on the rear of its shell. Oceanographers wondered why, and discovered a subtle, surprising glow from the vents themselves. The shrimp monitors this unexplained light to keep hot fluid from scalding it.

These are huge variations, as faces go. But here as everywhere, a small change can sometimes spur a destiny.

Why Have a Hairless Face?

We treasure smooth facial skin and can respond very badly to interruptions in it like acne and wrinkles. A particular grotesquerie is hair all over the face, the rare "werewolf" syndrome. Yet hair coats the faces of most mammals. We have a bare face, and this apparently trivial fact has shaped our very nature.

It goes back to primates. According to a tale in the *Popol Vuh*, the Mayan creation epic, the gods' first attempt to populate the earth resulted in people with "dry faces." The gods deemed these wooden specimens a first draft and tried to expunge them. Their descendants inhabit the jungles today: monkeys.

Monkeys are far from wooden, of course. They are quicksilver creatures, agile, social, ceaselessly achatter. And their "dry" or naked faces can stand out strikingly, isles of color amid fur. It is an innovation, for their immediate forebears, the prosimians like lemurs and lorises, have hairy faces.

What chased the pelt away? A big clue lies in the upper lip, where a second change occurs. In most mammals, the upper lip clings tightly to the gums. That's why no real cat will ever grin like Garfield, and why a title like *The Jaguar's Smile* implies fantasia.

But in monkeys the upper lip is free and moves about deftly. It lets the countenance take a plethora of shapes, and each can be a signal. The face thus grows more articulate. And since these signals must be visible, the fur withdrew. Our faces are bare so others can read them.

Prosimians show the alternative. They communicate mainly by odor. A ring-tailed lemur (*Lemur catta*), for example, will scent its tail and swirl it about to disperse messages. The aroma wafts out slowly and it dallies. The signal is a drone.

But the naked face is a dance of meaning. For instance, a monkey ready for fun will show a "play-face," a near-grin that displays the lower teeth. Others see it instantly, and a second expression can follow at once. Hence these animals convey far more every minute. It's like the difference between smoke signals and live video.

The earth abounds with social creatures like dogs and lions, and they too have face signals. But the smooth face greatly expanded the vocabulary, made messages clearer, subtler, and more varied. It hooked monkeys into a dense, rapid information web and led to supersocial creatures. Chimps console each other and play intricate political games, for instance, and we humans are virtuosos of cooperation. Our ability to gauge trust and work with others depends partly on the face, and it has let us farm, mine, and wire the earth, beat back predators and disease, and dwell in rich cities and suburbs. The hairless face was a first step to civilization.

The Great Resculpting

Few ideas jarred the nineteenth century quite like natural selection. Many thinkers felt an ape ancestry was impossible, even

insulting, given our broad minds and deep souls. In one notorious jibe, Bishop Samuel Wilberforce wondered whether evolutionist Thomas Huxley had descended from apes on his
grandfather's or grandmother's side.*

Yet suppose apes' faces looked like ours. Would the bishop
have ventured this jest? Would fundamentalists be quite so sure
Darwin was wrong if they could see the spitting image of themselves beyond the moat in a zoo, calmly stripping leaves off a
branch?

Our faces don't resemble those of apes or of any other animal, and it's one reason we've deemed ourselves so singular.
Indeed, the recent remake of the human countenance is the
most arresting part of its history, and unprecedented in evolution. For our faces stem partly from the products of our minds.
Each of us has a smart-face, bred of weapons, fire, and desire.

The steppingstones to humanity lie in Africa. They begin
with walking apes, like the famed fossil Lucy, who dwelt in
transitional woodlands between 5 and 1.3 million years ago.
Known as australopithecines, these creatures looked much like
chimps, with their swooping muzzle, chunky teeth, and wisp of
forehead.

Yet their upright stance demilitarized the face. A four-
legged animal's jaws and teeth reach forward, like a spearpoint,
and make the face a natural weapon. A wolf, for instance, lopes
along with fangs in front, and even chimps routinely nip their
rivals. On two feet, with head poised atop shoulders, the

---

*In his famous riposte, Huxley said, "If the question is put to me would I rather
have a miserable ape for a grandfather or a man highly endowed by nature and
possessed of great means and influence and yet who employs these faculties and
that influence for the mere purpose of introducing ridicule into a grave scientific
discussion, I unhesitatingly affirm my preference for the ape." But witnesses report
Huxley was white-hot with anger and failed to deliver the line effectively.

australopithecines not only lost this protective design, but left their whole bodies open to attack.

So how did they fend off the sabertooth cats and hyena packs of the transitional forest? Did they have a vicious kick, like the ostrich and kangaroo? Perhaps they had sharp claws, or wielded wooden sticks. In fact, they probably fled up trees. They were not pure bipeds. Their curled feet and apelike semicircular canals, or balance sensors, suggest they lived part-time in the boughs like chimps. They may have evolved a two-legged stance to cross the ground from tree to tree more quickly.*

The australopithecines lived for eons, but by 2.5 million years ago the globe was cooling and the current pattern of seesaw ice ages had commenced. Africa dried and the trees thinned out, and these creatures may have become easier marks as they raced between them. They dwindled away and a new animal appeared: *Homo habilis*, the first member of our own genus.

Here was a radically different creature. *Homo habilis* didn't scurry up trees. It was strictly two-footed. How did it stave off carnivores? It almost certainly used weapons. It could hurl rocks at them, but more significantly, it made stone tools. Some were sharp flakes that could have drawn blood from predators and cut tough hides, possibly letting *Homo habilis* dine on meat.

Homo habilis was novel in other ways. Its brain grew 50 percent, an astonishing development. And its face began drifting toward the human. Its forehead lifted a bit, its muzzle slimmed,

---

*Few disciplines are as confused and tumultuous as prehuman anthropology. Evidence is scant, new finds routinely incinerate accepted truths, and everyone, it seems, disagrees about everything. The account here owes much to Darwin and Steven Stanley, a paleobiologist at Johns Hopkins. Other scholars suggest upright posture evolved to let creatures roam greater territory, carry things, cool the body, or signal appeasement better (and thus reduce mortality). Several or even all may be correct.

and its teeth shrank, perhaps because they mattered less as weapons. Some had heavy browridges, which anchored the jaw muscles, and the first nub of a projecting nose.

From *Homo habilis* through *Homo erectus* to us, the four key changes occur: The face flattens. The forehead rises to house the ballooning brain. The nose juts out. And the chin appears. The first three commenced early, and the chin debuted almost yesterday.

The true human face appeared at the end of a deep ice age 130,000 years ago, with modern *Homo sapiens* in Africa. It differed strikingly from that of even the Neanderthals, our closest cousins. Indeed, when archeologists want to tell the latter from us, they look first to the face. The Neanderthals had bulging browridges; we just have eyebrows. They had moonlike skulls; ours more resemble short loaves of bread. They possessed long, narrow jaws and massive teeth they apparently used as a clamp. Their noses were great fleshy sodbreakers. They had cavernous eyesockets and virtually no chin. And, notably, they retained a modest muzzle. Our faces are flat.

The vanished muzzle may be the most beguiling evolutionary fact of all. A projecting mouth is essential equipment in almost all vertebrates, from pike to polar bears. It lets them snare, gnaw, and nip.* Yet we don't need it at all. As Darwin suggested in *The Descent of Man*, our brains made the muzzle obsolete.

A muzzle thrusts teeth outward so they can close like a trap, killing and wounding. Leopards maul prey and camels bite attackers. But our teeth lie within the skull and they make awkward weapons. We manufacture better ones instead, and our hands have evolved myriad grip positions to handle them and other tools.

---

*In most animals, the mouth also actively seizes food. Horses nibble grass and bears munch directly on kill. We, and most primates, use hands to bear food to our mouths.

We also gnaw food less, since we gained control over fire. Hearths first appear around 300,000 years ago, though they don't become common in archeological digs until around 40,000. Cooking softens food, reducing the need for strong jaws and teeth, and if we used fire more than the Neanderthals, it might also explain our loss of browridges.

And we rarely nip our fellows, as chimps do. We fling words instead and the right ones can sting, as witness the Wilberforce-Huxley exchange. They won't stop a grizzly, but they do very nicely with other people. If the muzzle lingered on for nipping, language could have obviated it.

These advances made the muzzle pointless. But it might still have persisted, a genetic free-rider like the appendix. It didn't, and archeologists have wondered why.

The reigning explanation invokes desire, and centers on the allure of the childlike face. We find babies winsome, since ancient folk with this trait paid more attention to their children, raised more healthy adults, and thus spread the gene that makes us google at infants. The attraction carried over to babyfaced grownups. They looked more appealing, reproduced more often, and passed on more babyface genes. The muzzle sank as if punctured and the face came to look infantile. The theory may be right. It's hard to test.

The individuals of 130,000 years ago were anatomical us. They had our foreheads, cheekbones, and flashing teeth, and if we could have looked them in the eye, we would have understood what we saw.

Double Star

"O! What a life is the eye! what a strange and inscrutable essence!" wrote Coleridge. Indeed, the eyes are far more than tools of sight, and we have just begun penetrating their glittery mysteries.

GENESIS

Nothing else shows thought like the eyes. They are the psychological center of the face, Pliny's "window to the soul," whose glow can speak intelligence and love. They are little pools of being, and they can bewitch us.

The eyes are as close as we get to seeing a mind. They can be dreamy, contemplative, vague. Elfride Swancourt's in Hardy's *A Pair of Blue Eyes* are "a misty and shady blue, that had no beginning or surface." Eyes can glint like laser pricks; one can "look daggers" at another. They can dart about like a trapped animal or twinkle with mirth. They can perform immelmanns of scorn. A look in the eye is human contact, and as Claude Lévi-Strauss found in India, can instantly spur beggars to solicit. Eyes can murmur sweet enticements; one can have "bedroom eyes." And when thought and feeling are absent, they can look hard as marbles, or even rubbery, like Popeye's in the Faulkner novel *Sanctuary* (1931).

Subjectively, we exist just behind our eyes, which form a transparent scrim on the world. We both oversee the outside and lie exposed to it. Hence the eyes are the most powerful and most intimate part of the face. In anger they flare, as if thought alone could scorch a target, while in shame people avert them, hiding the mind from view. And in delight and love the eyes sparkle, beckoning people to peer in. Indeed, lovers gaze into each other's eyes and feel a dizzy freefall.

The eyes seem alive, and when a razor slits an eye in *Un Chien Andalou* (1929), the audience always gasps. Photos of eye surgery are notoriously hard to look at. Some peoples have mutilated eyes from fear of their power. The Ainu of Japan dug a knife into the orbs of a slain bear to keep its spirit from seeking revenge, and the Parintintin Indians of South America ate the eyes of dead foes, to blind their ghosts. Images fare no better. Muhammad aimed first at the eyes when he attacked the idols in the Kaaba. During the Reformation, Dutch iconoclasts gouged at eyes in paintings, and when Easter Islanders toppled the basalt faces in their grim civil war, they methodically shattered the eyes.

For centuries many criminals believed murder victims retained an image of the killer's face on their retinas. Hence, after Frederick Guy Browne slew a constable on an English roadside in 1927, he leaned over and fired one shot into each of the dead man's eyes. Police caught him anyway.

The eye long seemed to mock evolution. How could such a splendid tool have appeared in stages? What good is half an eye? Yet the evolutionary stages are out in plain view. Protozoa have dotlike "eyes" or photoreceptors that register the mere presence of light. Limpet eyes have receded into pits. In abalones, the pit has almost closed over, forming a pinhole eye like a camera obscura. And squids, octopi, and most vertebrates have full camera eyes, with lenses that form sharp images. Using computers and cautious assumptions, biologists Dan Nilsson and Susanne Pelger have estimated that an animal could go from a flat skin-eye to a camera-lens eye in 364,000 generations, or in most cases less than 500,000 years.

We possess two orbs, like every other sighted vertebrate except the four-eyed fish. Monoculars like the bloody Cyclopeans of *The Odyssey* and the griffin-fighting Arimaspi in Herodotus live only in myth. Even primitive worms like the half-inch *Planaria*, which dwells under rocks and in streams, have paired eyes. Two eyes show an item from separate angles, so it appears against a slightly different background on each retina. The brain assesses this discrepancy and thus gauges distance, a trick called parallax. Two eyes also provide backup If some Odysseus drives a hot pike into one, we, unlike Polyphemus, have another.

The visible eye is about one-sixth of the entire ball, and its fascinating effect stems from its three interacting parts: white, iris, and pupil.

The white is part of the overall eyeball sheath, the sclera, which becomes transparent over the iris and pupil. Its gleam resonates with the teeth, and in a brilliant glance the two can seem to swap electricity.

Does it matter that it's white? Would blue suffice, or ochre? In fact, the ivory color is crucial. Since it contrasts with the darker iris and pupil, it highlights eye movements. If the sclera blended in with the iris, we would have trouble telling where people were looking and flail about socially.

Detecting gaze direction is a vital ability and our brains have special wiring for it, a weathervane for glance. It tells us whom individuals are looking at, focusing on. Hence if we see that a person is angry, we know whether he's menacing us or another. When we enter a group, this skill rapidly builds a social map, showing who is heeding whom. One scientist suggests this "attention structure" quickly clues us to hierarchy and fosters social coherence. It organizes us.

Our sensitivity to eyes yields other boons. Gaze implies one's next movements, and thus can signal purpose or desire. For instance, gorillas in the zoo will look to an object they want, then beseechingly toward a human. We say "with an eye to" and "with a view to" to denote goal, and the Zulu phrase *isa liwela umfela ugcwele*, "yearning reaches the impossible," means literally "the eye crosses a flooded river."

In fact, eye movement sends a constant stream of messages, and it may lie at the core of the striking eloquence of the eyes, as we'll see. But without a backdrop like the white, this language would simply elude us.

Between white and pupil lies the iris, a chromatic ring. It is not one flat hue, but a riot of spots, wedges, and spokes. Its color also changes from pupil to perimeter. Each iris pattern is unique, and experimental ATMs are already using them to identify customers.

The jumble of hues in the iris can make eye color a matter of opinion. Novelists have shamelessly exploited this latitude, ballooning interstitial tints and using the results in a subtle color-code of character. For instance, yellow eyes invoke the feral. The Phantom of the Opera and Frankenstein's monster have yellow eyes, and Rosemary's baby has orbs of golden yellow, whites and all, with black-slit pupils like a cat's. Gold hints at greed and allure. Balzac endowed the miser Grandet with such eyes, and the bisexual houri in his *The Girl with the Golden Eyes* (1835) has eyes of "living gold, brooding gold, amorous gold." Gray eyes suggest inscrutability, and bedeck such fairly opaque characters as Homer's Athena, Flem Snopes, Lolita, Buck Mulligan, and Bartleby.

Colored contact lenses opened a Pandora's box, enabling iridescent eyes, mirror eyes, square pupils, written messages over the entire cornea. They allowed designer eyes and FX specialists exploited them. For instance, Linda Blair's vivid green eyes in *The Exorcist* lent an extra jolt to her possession. Today, computers can make actors' irises change color and even show little films.

The iris is actually a pair of muscles, the most beautiful ones in the body. They work like a camera diaphragm to change the aperture of the pupil, letting more or less light reach the retina. One set of fibers radiates out from the pupil. If you enter a dark theater, these marionette strings pull the pupil open. The other fibers coil round the pupil like a noose. Return to the glare of sunlight and they contract, shrinking it. Without the iris, we would often be blind.

We come at last to the black heart of the eye, the pupil. It is the object of the iris's embrace and the opening onto the wonders of the retina. One out of every five people has pupils of different diameter, which can actually change size independently and alter the balance in a few hours. Blue-eyed individuals possess larger ones, on average, than brown-eyed. The pupil is not uniformly circular in humans, and in animals it varies from the keyhole slit of the lemon shark to the horizontal blob of sheep or cattle. Shape doesn't matter. They can all control the amount of light entering the eye.

In people, the pupils take on an extra role. They are profoundly expressive. These obsidian disks widen not only in dim light, but before an image that excites us, as shrewd poker players and bargainers know. Men's pupils dilate when they look at photos of sharks and female nudes, women's when they see pictures of babies, mothers with babies, and male nudes. The pupils mirror our level of awareness overall. Fear, surprise, joy, anxiety, loud noise, and even music will expand them, and boredom and drowsiness shrink them.

We tend to like those who care about us, so big pupils attract us. Researchers showed men pairs of photos of women identical in every way except that retouchers had enlarged the pupils of one, and found men preferred her but couldn't say why. Hence dark eyes seem romantic. Rochester and Hester Prynne have deep black eyes, as does Lotte in *The Sorrows of Young Werther* (1774). Our pupils reach peak size in adolescence, almost certainly as a lure in love, then slowly contract till age sixty.

The eye dances with profound little messages, the source of its life. Movement and pupil size are two of these signals, and a third is even subtler.

Cutting Room of the Mind

Even the busiest attorney or executive loses about twenty-three minutes in every waking day. They simply vanish and we are oblivious to it, as if spellbound. The time trickles away in 14,000 tiny gaps, which the brain edits out of our perception. They are blinks.

Blinks are like the spy who lived next door. They seem utterly ordinary yet hold many secrets.

They recall our aquatic origin. When the first tetrapods crept onto land some 370 to 360 million years ago, they faced the peril of desiccation. In water, animals never dried out; on land, the sun stole their very substance. Early amphibians slipped in and out of water often and reptiles evolved near-watertight scales. But eyes contact air directly, so to keep them wet, creatures developed eyelids with tear glands. Each blink is a little dunk in the primeval sea.

The eyelid is the tool of the blink. Cicero called it "exquisitely designed by nature," and he could not have been more correct. For instance, the eyelid blocks out the world when we want to sleep. But since it is just one millimeter deep, the thinnest skin on the body, it is slightly translucent, so sunrise or sudden light can wake us up. The eyelid helps register alarms.

Above the outer corner of the eyelid lurk the tear glands, which emit a fluid that the lids sweep over the eyeball. These tears are not just water. They are part of the circulatory system. The cornea must be transparent so we can see through it, and hence it has no blood vessels. Tears bear oxygen to it and keep it alive. They also contain chemicals that kill bacteria and proteins that smooth the eye surface and trap debris. A quarter of tear fluid evaporates, and the rest seeps down into the nasal passages and keeps them moist. That's why crying makes us sniffle.

Eyelashes highlight the blink. They are a moving palisade, holding insects and other hazards at bay, and indeed blinking is a reflex to any threat to the eye. The lashes perform the same task as the hairs in the nose and ears, but they live in Elysium compared to their troglodyte cousins. They are the eyes' attendant graces. Emma Bovary's brilliant brown eyes, "her real beauty," seem black because of her lashes. The starburst pattern of lashes draws attention to the eye, and flirts flutter them to gain even more. Eyelashes are erotic in other ways. In *Remembrance of Things Past*, the narrator and Albertine entwine eyelashes in bed. Malinowski says Trobriand Islanders bit off eyelashes in love-making, an act they called *mitakuku*.

We say "in the blink of an eye" to mean "instantly," but objectively the comparison falters. The average blink lasts a third of a second, and the lid covers the pupil for a sixth of a second, during which we are blind. (Deliberate blinks take longer and no one knows why.) This gap is far from an instant, since we can detect flickers as brief as 1/300th second in a lantern. In fact, a sixth of a second is long enough for an attacker to catch us off guard, and biologists like George Williams have wondered why we always blink both eyes at once. If we alternated blinks, we could see the world all the time.

On the other hand, subjectively the comparison is almost too good. Since the brain kills awareness of normal blinks and knits the world into one flowing vision, a blink seems to take no time at all. It feels shorter than an instant.

Blinking is like breathing. We do it automatically, all day long, about 15 times a minute. Dry spots begin to pepper the eyeball after 15 to 45 blinkless seconds. Slow-motion photography reveals the career of a blink. The lids shut like a zipper, traveling inward from the outer edge to spread the tear fluid. They close twice as fast as they reopen, like a coquette turning her

333333333333

head and slowly looking back. Surprisingly, we rarely complete a blink, unless we're blinking deliberately. It doesn't seem to matter.

Pliny the Elder says that of the 20,000 gladiators in Caligula's training school, only two did not blink when facing a threat. They were thus unbeatable. In folklore, the blink has long suggested broken concentration. Whoever "blinks first" has lost focus and nerve. We say "without batting an eye" to mean "apparently unperturbed," and a "blinkard" is a dimwit.

Science supports the folklore, in essence. When we look or even listen intently, our eyes stay open, as if to suck in every mote of information. But when our attention strays, we blink. We blink less while we're reading sentences, and in bursts in between. One experimenter found that his best readers kept their eyes open for an entire page, and released a flurry of blinks when turning it. Drivers blink as they turn to glance at the speedometer, and again after they've assessed the velocity. We blink less when tracking an object or working a maze and more just afterward.

On the other hand, as we grow bored or fatigued, we blink more. If we drive a car or read for a long period, the blink rate rises. In one study, people were blinking 6.9 times per minute when they started reading and 11.0 times per minute four hours later. As concentration wavers, the eyelids dance.

Speaking also boosts the blink rate. For instance, doing mental arithmetic doesn't change it, but if one verbalizes the steps, it jumps. Reciting the alphabet silently slows the rate; reciting it aloud quickens it. We blink less often while listening to a question and formulating an answer, and more while delivering it. Blink frequency also rises during cross-examination on the witness stand, or even just talking with a friend. We may blink more when speaking because we are absorbing less information or, some scientists think, simply because we are moving the tongue.

John Huston likened blinks to cuts in movies, since we blink when shifting our gaze from one spot to another. Walter Murch, editor of such films as *Apocalypse Now* and *The English Patient*, goes further. Natural blinks can almost dictate cuts, he says. Where a careful listener blinks, a film editor can probably cut. Skilled actors also blink at good cut points. Blinks help us make sense of a continuous world by dividing it into little chapters, he feels, and ultimately, the film editor is blinking for the audience, cuing it to discrete thoughts or acts.

Blinks communicate. Actor Michael Caine eschews blinks in close-ups, believing they lessen intensity, and for years actually practiced not blinking. Murch suggests that blinks show others when we have grasped an idea and thus subtly coordinate conversation. Bad actors, he notes, don't think their characters' thoughts and so blink at the wrong moments. So do politicians. Such miscues disturb the rhythm and we pick them up, he says. We sense unnaturalness and often assume such people are lying. He may be right. Blink science remains in its nonage.

Sphinx

An obvious fact is as plain as the nose on your face. Yet aside from its blatant existence, few things are plain about the nose. It is esthetically deceptive, symbolically bipolar, physically protean, and even semi-secessionist. It has lodged right in the middle of the face, and there it flings riddles at us.

What kind of nose is most attractive? In *Little Women*, Amy feels her flat nose has been the great misfortune of her life and puts a clothespin on it to try to extend it. In Rabelais, Friar John says firm breasts in wet-nurses halt nasal growth and give children

The Nose.

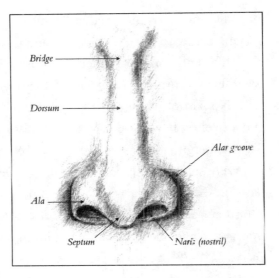

ugly snubs. Renaissance theorist of beauty Agnolo Firenzuola (1493–1543) deemed a turned-up nose unsightly, and Malinowski reports the Trobriand Islanders felt a flat nose was unattractive and limited the potential for romance.

But others idealize the retroussé ("turned up") nose. Thackeray gave Becky Sharp a snub. Dickens attributed criticism of it to envy, and Marilyn Monroe practiced holding her upper lip down when she smiled, to make her nose seem smaller.

Scientists have now resolved this strange controversy: Men overwhelmingly prefer a small nose in women. Indeed, when cartoonists omit the nose of a pretty woman, men fill in with a snub.

The nose is a paradox in psychoanalysis. Freudians classically deem it a symbol of the male sex organs, and in the Slawkenbergius tale in *Tristram Shandy* a horseman visits the Promontory of Noses and gets a near-obscene one. Yet psychoanalysts also say the nose emblemizes the female genitals, though not as frequently. It is both protrusion and opening.

The nose has bemused anatomists. Even today, they give varying names to the same muscles and cartilages, draw its muscles differently in established textbooks, and disagree about whether one muscle, *M. nasalis*, widens or shrinks the nostril.

No part of the facial flesh feels as autonomous as the nose, and novelists have played extravagantly on this sense. In the Slawkenbergius tale, the horseman replaces his nose as casually as if it were spectacles. Judge Whimplewopper in Ishmael Reed's *The Free-Lance Pallbearers* (1967) has a nose so long he has to rest it on a purple satin pillow. Fans of it throng the corridors outside court and seek the nose's opinions. In *The Adventures of Pinocchio* (1883), a nose tattles on its owner. The device occupies just one percent of the book, yet it resonates, capturing our fear that, when we lie, our faces may betray us.

Nikolai Gogol (1809–1852) brought noses to full-fledged life. In "The Diary of a Madman," the title character says noses have built a civilization on the lunar surface. "That's why we can't see our own noses: they are all on the moon." Platon Kovalyov's proboscis in "The Nose" (1834) is the most interesting specimen in literature. When a barber slices it off, it gains freedom and carves out a career for itself in St. Petersburg. Kovalyov spots it in a gold-braided uniform and begs an audience. It is proud, in the way of noses, and dismisses him curtly.

Indeed, a hubris attaches to the nose. An arrogant person has "his nose in the air" or "turns his nose up." In Japanese, *hanataka*, "high nose," means "proud." The obsolete term "nose-wise" meant "clever, in one's own opinion." Samuel Johnson defined "to nose" as "to look big, to bluster." We "thumb our noses" at others, putting a thumb on the nose and wiggling the extended fingers in derision. Pride's link to the nose reflects a biological fact: We wrinkle it automatically to show disgust.

Yet the nose can also signal humility. It can be passive, a kind of handle — not a sign of respect for its owner. We can tweak a person's nose, or pull it. A "nose of wax" is a weak person, easily manipulated. The Serbo-Croatian phrase *vuci za nosa*, literally "to drag by the nose," means "to make a fool of." In Gogol's "The Quarrel of the Two Ivans," a woman grabs Ivan Nikiforovich's nose and leads him around like a poodle. "Is that all our noses are good for?" the author laments.

The nose is the most variable part of the face. It can be snub, ski-slope, bulbous, bent like a boomerang. It can be aquiline ("curved like an eagle's beak"), straight, Roman ("having a prominent, slightly aquiline bridge"). The classic "English" nose, like Henry VIII's, is straight with a delicate camber. The nose can be flat and wide, or splayed out and close to the skin. It can be long, high, and narrow, like a blade.*

Comedians like Jimmy "The Schnozz" Durante and Phyllis Diller gained fame with their noses. In *The Bank Dick*, a little boy points at W. C. Fields and laughs, "Look at the man's funny nose!" His mother scolds him: "Mustn't make fun of the man's nose, dear. You'd like to have a nose that big full of nickels, wouldn't you?"

Fields suffered from rosacea, a peculiar disease that makes the nose a scarlet flare. It exaggerates the normal flushing of the face and ultimately swells the capillaries, causing constant red-

---

*Da Vinci held there were ten types of noses in profile: "straight, bulbous, hollow, prominent either above or below the center, aquiline, regular, simian, round, and pointed." In full-face, he said there were 256, derived by combining six variables: 1) tip (broad or narrow), 2) base (broad or narrow), 3) middle (thick or thin), 4) nostrils (broad or narrow), 5) nostrils (high or low), and 6) openings (visible or hidden by the tip). Of course, this binary scheme ignores the myriad values in between, which inflate the figure beyond reckoning.

ness. The extra blood can make the skin overgrow and dot the nose with ugly pustules. Alcohol flushes the face and worsens the ailment, which came to be known as "drinker's nose" or "grog blossoms." J. Pierpont Morgan (1837–1913), another famous victim, offered $100,000 to anyone discovering its cause. Scientists have yet to collect, though they have linked it to the wicked-looking mite *Demodex follicularum*, and one new theory points to *Helicobacter pylori*, a bacterium that causes ulcers. Some physicians believe Bill Clinton has rosacea.

Sun also reddens the nose, and faster than any other part of the face. Because the nose protrudes, ultraviolet rays can strike it from almost any angle, especially the side, while the forehead and other surfaces gain respite. Like a salient in a battlefront, the nose takes more bombardment.

Probably no literary figure has described the nose more stirringly than Cyrano de Bergerac. Dramatist Edmond Rostand (1868–1918) once wrote billets doux for a friend in school, though apparently not to a woman he loved. Yet he clearly had a sense for Cyrano's plight. "A large nose is the sign manifest of such a man as I am — courteous, witty, liberal in opinions, fired with courage," says Cyrano at one moment, and at the next he bemoans his nose as a curse. Any extreme facial feature can harm looks, but the nose is especially hazardous. Oddly, Cyrano, for all his brilliance, missed the one insight that could have saved him: Roxane didn't care about his nose.

Why do noses jut out at all? Coleridge suggested it was so people could take snuff. The nose puzzled Firenzuola, who listed uses for it — breathing, smelling, and purging — then shrugged and said it seemed mainly an ornament.

In fact, our facial prow is zoologically bizarre. Galapagos

tortoises have just two holes in their heads, and they are typical. The nares of fish and lizards don't project. Neither do those of gorillas or chimps. Among living primates, only the proboscis monkey has a protruding nose.

Perhaps it is an odor canopy. Since we are vertical and flat-faced, the nose may gather aromas rising from below, so we can better assay food. We mainly sniff viands — unlike most animals, we rarely use scent for prey, social signals, or other items at a distance — so mouth-facing nostrils may make evolutionary sense. On the other hand, we don't know how much good this configuration actually does, and it first appeared on the slope-faced *Homo habilis*, where it would have done less.

Maybe it began as an anteroom for breath. It certainly serves this function today. As air passes up the nostrils, it picks up heat and moisture from the mucous membranes, so it won't chill or dry the lungs. Hence tropical peoples have smaller noses than arctic ones, and the cold-dwelling Neanderthals possessed gigantic noses. However, almost every warm-blooded animal performs this task inside the head, in convoluted passages called *turbinals*. We have turbinals too, and the question here is why they should have partly migrated outside the skull, where they are more vulnerable.*

Could the nose be just a lowly servant of the eye? That's the theory of psychologist T. G. R. Bower, who observes that every animal with panoramic vision has some projection of itself that interrupts its view. For instance, chimps have muzzles, dogs have snouts, and the owl has a nasal tuft that covers 30 degrees

---

*Turbinals do more than precondition air. They keep us from dehydrating. They collect moisture from exhalations, and without them mammals would lose 75 percent of their daily water intake.

of its visual field. Bower argues that human eyes compare the world against the nose. The nose is always visible, so it helps us position objects and tell whether they or we are moving. This idea is so delectable it's hard not to root for it.

In 1960 biologist Alistair Hardy suggested an even more radical notion, one writer Elaine Morgan has since expanded on. It has become one of the most ridiculed and interesting ideas in prehuman anthropology: the aquatic-ape theory. It holds that we spent some recent evolutionary time partly in water, perhaps to elude lions and other carnivores. During this period our bodies changed. We lost most of our body hair, since it no longer kept us warm, and instead developed a fat layer under our skin, like dolphins, seals, and other marine animals.

Water altered us in many other ways, the theory contends. For instance, it reduced our sense of smell, and indeed among mammals only dolphins and whales have worse olfaction than we do. The following table, taken from Morgan, compares features in apes, people, and aquatic species, such as hippos, dolphins, and penguins. (A Yes in the latter column indicates that several animals share the trait.)

What good is a projecting nose to a water-dweller? Many primates are good swimmers, but the champ is the proboscis monkey (*Nasalis larvatus*) of Borneo. The proboscis swims underwater and dives from 50 feet, higher than any Olympic platform. It has a fantastic nose, a sock of flesh that in males actually overhangs the mouth. This nose protects the monkey's nasal cavity from headlong rushes of water. So does ours. At the beach it deflects waves, and when we swim it diverts the slipstream flowing past. If we were noseless like gorillas, we would constantly face the surprise of water-choked turbinals.

The aquatic-ape theory has surface appeal, yet so far most scientists have ignored it. It is hard to see how some human

	Apes	Humans	Aquatic
Prominent nose	No	Yes	Yes
Sparse body hair	No	Yes	Yes
Streamlined body hair	No	Yes	Yes
Subcutaneous fat	No	Yes	Yes
Tears	No	Yes	Yes
Baby's weight heavy compared to mother's	No	Yes	Yes
Midwives	No	Yes	Yes
Sense of smell	Good	Poor	None
Diving reflex	No	Yes	Yes
Automatic swimming in babies	No?	Yes	Yes
High brain–body ratio	No	Yes	Yes
Conditionable vocalization	No	Yes	Yes

features, like babies' ability to survive for an hour underwater, could have arisen without a watery environment. Yet until this theory survives an enfilade of scientific criticism, its merit will remain unclear.

The structure beneath the nose has a more obvious role.

The Primeval Feature

If we view the face as a geography, it is two long forests, a pair of multicolored sunken lakes, a Gibraltar-like peak, and an abyssal pit. The pit is the most dramatic item on the map, and it is of course the mouth.

The mouth is the oldest part of the face, the gateway for food, drink, and at times air. It is primal, essential. Even the one-celled paramecium has a mouth. We can survive outside a hospital without eyes, nose, or any other facial feature, but without a mouth we starve. Pliny cites learned authority for the existence of the Astomi in Pakistan, who lack mouths and live on the aroma of roots, flowers, and apples. In fact, some simple creatures do lack a mouth and absorb food through their "skin." But it is vital for any kind of digestive tract.

The mouth is the first facial organ to form. It arises in embryos soon after fertilization, in the process called gastrulation. A dent appears in the spherical embryo and deepens, pulling in the cells that will make up the muscles and inner organs, and forming the gut. When this tube reaches the far end of the embryo, it breaches the cell wall and creates the mouth.

This aperture is the body's main entrypoint and guards hedge it about. Before swallowing food we test odor, taste, texture, temperature, shape, and irritability, and we have a gag reflex to stop it at the last minute. Our emotions also patrol this ground. For instance, saliva feels normal in the mouth, but spit some in a glass of water and it becomes instantly repulsive. Yet it is the same saliva.

The mouth is the most plastic part of the countenance. Eyes move constantly but subtly, in a delicate shimmy of jumps and tremors. The mouth often rests, but once in motion, it can sigh, yawn, smile, laugh, drop open, pout, tremble, and tighten. When we speak, the mouth moves in seemingly endless ways: widening, opening, closing, puckering, protruding, retreating. And if we like, we can twist the mouth into weird topologies. It is the contortionist of the face.

Our mouth sets us apart from other creatures in several ways, but the most striking is its narrowness. Human mouths are usually no wider than the span between the pupils, a surprise to most beginning portrait artists. Some fish like carp also have tiny mouths, which they use for suction feeding, but they and we stand almost alone. Indeed, next to us even chimps have huge maws, running hairline to hairline, and horse and alligator faces are almost all mouth. Why did ours shrink?

First, because it could. Muzzles need wraparound mouths, but ours is free from this constraint and has evolved to different ends. One is protection. Since we don't use our teeth as weapons, the mouth need not be wide, and a small one is safer from contaminants. Another is probably expression. The face muscles can control small lips more deftly, enhancing our array of smiles and grimaces. A third may be language. Sounds like "oh" and "w," for instance, require the lips to form an O.

Intriguingly, the large mouth has rarely been a sign of beauty, but the small mouth has. In Victorian times, especially, a tiny mouth was dainty. Trollope's women often had such a mouth, as did Dickens's Little Nell.

The mouth is the home of two notable facial transients: the teeth and the tongue.

Teeth are performers, beaming into view as we speak and smile. They give the face delicious volatility and can dazzle us with their ivory gleam. "Toothsome" describes a form of beauty and radiant teeth seem the reward of a smile. Firenzuola said the teeth impart "so much charm to a pleasant face that, without them, sweetness does not seem to reside too willingly upon it." Their secret is less sublime: saliva. It coats the teeth with water, so they gleam in the light.

Teeth last longer than any other part of the face—not surprisingly, since enamel is the hardest biological substance known.

South American cannibals made necklaces from the teeth of devoured enemies. States have warred to own a tooth of the Buddha, who preached against ownership. The most precious tooth on earth today came from the Buddha, we are told, and resides in a gold vessel in Kandy, Sri Lanka. Gary Snyder, who viewed the plaster cast of it, said it was two inches long.

Modern teeth arose with the advent of jawed fish around 440 million years ago. Only jaws can grip, manipulate, cleanly cut, and grind prey. The first jawed fish, the sardine-sized acanthodian, prowled the seas until 260 million years ago. Its teeth were tiny stilettos like piranhas', but teeth quickly radiated into cones, blades, and crushers.

Teeth reach their apex in mammals. Most other animals use them mainly to seize prey. But mammal teeth can shear, crush, and grind food. They allow chewing, a notable power. Chewing speeds digestion and puts more of the natural world on the menu. It also leads to a dextrous tongue, for nudging food into position, and to polymorphic teeth.

Mammals boast a more complex array of teeth than other animals, an intricate palisade of incisors, canines, premolars, and molars. The incisors are the blades, cutting bites from carrots and apples. Canines, our relic fangs, have deep roots and handle tough foods like jerky. The premolars and molars crush food into pulp for the stomach. Behind the molars lurk wisdom teeth, which often never emerge. (They may in fact be evolving away, since with our flat faces we need fewer teeth.) Teeth have diversified so much in mammals that a paleontologist can often identify a genus and even a species from a single tooth.

People fiddle with the face and have altered every part of it, even, with laser surgery, the eyeball. They have throughout history improved on teeth. Pre-Columbian Indians of Mexico added inlays of jadeite, pyrite, gold, and turquoise. Tribes have notched, grooved, scratched, perforated, and sculpted their teeth, and in Mesoamerica alone, dental anthropologists have identified 59 different types of tooth mutilation. The Baule of Ivory Coast often remove a diagonal half of each front incisor, creating a black triangle when they smile. The Kadars and Malavedans of southern India file their teeth into points, as do the Tiv of Nigeria. Tiv soldiers stationed in Cairo found, to their delight, that Egyptians thought they were cannibals, and they quickly learned that fierce faces in the bazaar won them breathtaking bargains.

Dracula has sharp white teeth that extend over his lips — fangs, essentially. But most people prize well-ordered, soldierly teeth. Precontact Inuit and aboriginal Taiwanese had perfectly aligned teeth, but the modern diet of soft, refined foods has jumbled them. Etruscan girls wore braces as early as 700 B.C., and today orthodontic treatment is the greatest precollege expense most parents lay out for their children. Diet also fosters decay, and dentists plug caries and replace teeth that fall out. These are the most common repairs we make on the face.

We speak with the tongue and a talkative person is a "tongue-pad." Yet this odd and agile organ rarely emerges onto the face, and its appearance is often aggressive. We stick tongues out to insult people, a gesture that smacks of the child. The classic Medusa pokes out her tongue, and the Maoris howled their unnerving war cries with tongue far out. Indeed, they carved knives with a face as hilt and a tongue as glinting blade.

A tongue in the cheek is more ambiguous. Smollett's Roderick Random says, "I signified my contempt of him, by thrusting my tongue in my cheek." But today "tongue-in-cheek" means ironically, not to be taken seriously. In modern

Western society, a real tongue nestled in the cheek suggests hesitancy, pondering.

The tongue can express other feelings. Run lusciously across the lips, it is a classic sexual invitation. A slightly protruding tongue can imply uncertainty. One turned downward suggests bewilderment, and most of us have seen the gesture of goofy delight—wide grin, tongue down, eyebrows raised—signifying surprise achievement: "I don't know *how* I did that!"

Fish lack tongues. They commonly seize food by sucking water into their mouths, a trick that utterly fails in air, so when fish first ventured onto land, they sprouted tongues. Old World chameleons can dart their tongues out more than a body length to snare insects. Most birds have tiny tongues, but the flamingo has a thick, esculent one, and Roman emperors Elagabalus and Vitellius fed guests heaping bowls of them.

We think of the tongue as a single muscle, but it is actually a bundle of them which can shorten, lengthen, and widen it. Muscles outside the tongue also tug it, so it changes shape deftly, a key fact for speech. Some people can roll their tongue into a tube and others can't. It's a genetic variation of baffling purpose.

The flare-like papillae on the tongue contain most of our 10,000 taste buds, though these sense organs also dot the palate, larynx, and pharynx. Most register not just one taste, but several, and thus send information in polyphony to the brain.

The tongue helps us swallow, and we do so about nine times a minute while eating, once a minute while not eating. The latter act is unconscious and salutary. Saliva coats many microorganisms with mucus, and swallowing takes them down to the stomach, where gastric acids annihilate them. Billions of other bacteria inhabit the mouth, a large proportion on the tongue. Most are benign, and many form one more rampart of the body's defense.

Outside, the mouth is a little duchy, like the eye, and many vassal features lie in its sphere. Most notable are the philtrum, nasolabial folds, chin, and lips.

The philtrum is the shallow vale between the nose and upper lip. It rarely attracts attention on its own, but where its two ridges touch the mouth, the lip rises to meet them. In between, it dips slightly. The philtrum thus fathers the graceful notch in the upper lip.

The nasolabial folds flank the mouth. These two creases slant out tentlike from the wings of the nose down past the lips, bounding the cheek. Age deepens these pleasant lines into character marks. Since they magnify smiles and other mouth expressions, they are key signals of mind-state. Yet not one person in a thousand can name them.

The nasolabial folds vanish in a paralyzed face, but not in a corpse. Why? In our daily life nerve impulses pull the folds up and back somewhat. Paralysis halts these signals and the skin turns smooth. Yet after death the muscles freeze in slight contraction, and the lines live on.

The chin is unique to humans. Not even Neanderthals had one. It appeared around 130,000 years ago, with the first anatomically modern people, and it remains a puzzle.

It is not utterly pointless. It seems to aid mastication, though its arrival coincides with no known shift in diet or subsistence pattern. And it clearly sets faces off and helps us lock onto them better. As hair frames the top half of the face and jawline the sides, the shadow of the chin marks the bottom. In profile, its jut signals the base, like a serif in typography. Without a chin, the lower face would merge with the neck, and face shape would blur.

Such uses don't necessarily explain its origin. Indeed, it may be inadvertent. Some believe it simply appeared when the muzzle retreated, much as islands emerge when sea level drops in an ice age. The mouth flattened faster, leaving a bony knob that recalls our nipping and gnawing days.

Darwin observed that relics like the appendix take more diverse forms than other organs. They have lost their utility, so most variations in them are not harmful and persist. If the chin is actually a remnant of the muzzle, it should come in many shapes and sizes, and in fact it does: soft, cleft, jutting, dimpled, long. Ironically, this variety has become a use in itself, multiplying the range of possible faces and distinguishing us.

The chin is crucial to looks. A slight chin is unattractive, especially in men, and hence another theory involves sexual selection. In males, the chin grows during puberty in response to testosterone, so larger chins suggest its greater presence. Since testosterone weakens the immune system, its excess is an "honest" signal of good health in men, advertising their ability to resist disease. Hence women may have evolved to like large chins. George Williams suggests that, today, male chins may be developing into sexually selected structures like the antlers of Irish elk, and in fact the average size of chins has grown over the last 200 generations. They are expanding, not retreating.

At rest, the mouth is pure lips. Lips are the spice of the face, twin ruddy bulges separated by the dark line of the mouth, like a pair of cushions. They can be thick or narrow, bulging or inswept. At rest they can seem to pout, frown, or smile, and some lips form a "cupid's bow." All are a transition zone between the dry skin of the face and the moist mucous membrane of the inner mouth. Their surface is thin enough to reveal blood below, so lips look dark.

Lips are border guards. They excel at distinguishing foreign objects, and can detect a single hair in our food. They lie in the forefront of our oral defenses.

But lips also serve subtler ends. For instance, they cue us to speech. We read them subliminally, which is why out-of-sync dubbing in foreign films bothers us. Lips are especially helpful in borderline cases. In one study, people recognized 23 percent of sentences uttered in noise, but 65 percent when they could also see the speaker's face. Noise has bedeviled speech recognition technology, and the face could help cut through it. Says Dominic Massaro of U.C. Santa Cruz, "You can go from chance to perfect by just having the face." Massaro is seeking to supply the deaf with glasses that distinguish visually similar sounds, like *ma* and *ba,* by flashing different colors.

Indeed, we rely so heavily on lip signals that they breed the strange effect called the McGurk illusion: People presented with the sight of one mouthed phoneme and the sound of another typically hear it as a blend of the two. Visual *ga* and aural *ba,* for instance, yield *da.* We are hearing partly with our eyes.

Ventriloquists speak liplessly. The trick is ancient. Inuit magi used it, and Zulu shamans made warnings seem to issue from the wattles of their huts. Greek and Roman ventriloquists claimed divine spirits spoke from their stomachs — hence the name, which means "belly-talking." In the sixteenth century, belly-speaker Elizabeth Barton, the "Holy Maid of Kent," drew thousands of pilgrims to hear her prophecies against the second marriage of Henry VIII. He had her hanged at Tyburn in 1534. In rural China, female ventriloquists sometimes spoke through little dolls on their stomachs, but the ventriloquist's dummy did not become common until the latter nineteenth century.

"If a man is not born a ventriloquist, he will never become one," said Victorian ventriloquist Walter Cole. Such claims swathed him and other adepts in a useful mystique, but in fact people learn the art, as they learn magic and juggling. The illusion has two parts. First, performers must create a second voice. Edgar Bergen must sound like Mortimer Snerd. To do so, they hump up the rear of the tongue, as if uttering an *ng*, and force part of the vocal tone through the nose. Second, they must speak with still lips. The main hurdle is producing *b*, *p*, and *m* inside the mouth. Ventriloquists make the *b* and *p* by placing the tip of the tongue on the front teeth, and the *m* by touching the rear of the tongue to the roof of the mouth. These feats demand the patience of Demosthenes.

The lips also enhance facial expression. They enlarge the rim of the mouth and thus highlight smiles, sneers, and gapes. We pull the lips tightly in to conceal expression, much as we put a hand over the mouth.

And, most obviously, lips give sensual delight. They are twin pleasure puffs, rich with touch sensors. Lips are part endoderm, like the lining of the gut, and their boundary with the skin or exoderm forms a line between inner and outer self. It means that, when we kiss, our inner selves touch.

An Anatomy of Kissing

A kiss is not just a kiss. Indeed, it is a medium more than a single message. The Romans noted three kisses: of friendship (*oscula*), love (*basia*), and passion (*suavia*). The Talmudic rabbis identified a more formal trio: kisses of greeting, leavetaking, and respect. Like love, kisses can both exalt and degrade. They can signify treachery, as the Mafia's kiss of death, and abasement, as in "The Miller's Tale." They can be ironic. Mata Hari blew a kiss to her executioners before they shot her.

The kiss of love is the core. Byron called it a "heartquake," and it can make the lips more eloquent than any speech. Heine writes, "Yet could I kiss thee, O my soul, then straightaway I should be made whole." A kiss can pour love from lips to lips, two receptacles filling each other.

A love kiss fuses. Hence the first kiss signals new intimacy in a relationship, and can even begin it. The kiss-as-union became a staple of Renaissance poetry, and one of its masters was Johannes Secundus (1511–1535), whose lovers kiss eternally, swoon to near-death, diffuse their souls into each other's bodies. This merger is common too in medieval art. We see it in Giotto's *Meeting of Anne and Joachim*, where the two seem to form a single mass, and in his *Kiss of Judas*, where a cloak further unites the traitor with Christ.

The two most famous sculpted kisses show very different sides of love. In Auguste Rodin's *The Kiss* (1898), two naked lovers coil into each other. The woman wraps an arm sensually around the man's neck, while he touches her bare hip. She kisses him from below as she falls slowly, deliciously to the supine. The work is alive with dreamy whirl, an erotic vertigo. It somehow inspired Constantin Brancusi's *The Kiss* (c. 1910), in which two blocklike humanoids press flatly together and lace arms around necks. Their faces almost disappear into each other, and we see them mainly in profile. The sheer awkwardness of this embrace — their mutual dependence, their inability to fully grasp each other — makes it poignant. Fittingly, Brancusi actually carved it from a single stone.

Distance is no obstacle to the kiss of love. People kiss images of their lovers: drawings, photos in lockets, freeze-frames on tape. They kiss snippets of hair, letters, lovers' possessions, tombstones — any link to the person. They pucker lips and release them, to send kisses through the air, whether they see each other or not. They fountain up the lips an anticipated kiss, a request. The ancients drank out of goblets at the spot their lovers' lips had just touched, a kiss through time.

The kiss of passion, the deep or French kiss, makes two mouths into one. It has a splendid history as an emblem of animal

Auguste Rodin, *The Kiss.*
Courtesy: Giraudon/Art
Resource, NY.

lust, but of course it is also an act of tenderness and intimacy. The lovers' tongues caress each other, dance about the teeth and inner cheeks, bathe in each other's oral fluids. This kiss merges inner seas and resets the bounds of self.

Even deep kisses are tasteless — usually — yet they all have intense metaphorical flavor. Almost universally, they are sweet. They are honey, nectar, liqueur. Renaissance swains described their ladies' "mouths full of sugar and ambergris." The *Song of Songs* says, "Thy lips drip as the honeycomb, my spouse: Honey and milk are under thy tongue." In Nabokov's *Invitation to a Beheading*, Marthe's kisses taste like wild strawberries. Jimmie Rodgers sang about kisses sweeter than wine.

Kisses can be flames. The lover of the Persian poet Hāfez (1320–1389) worries his kisses "will char her delicate lips."

Constantin Brancusi,
The Kiss. Courtesy:
Philadelphia Museum of Art.

They can pass an electric thrill, and the touch of a kiss can linger in the mind for years.

They can be almost inaudible, like silk sheets rustling, or soft as a footstep in sand. In the final parting, air rushes in to separate the lovers, so a kiss can pop like a bubble or have the suck of a bottle opening. Long, luxurious smooches can end in a sound like panes of glass rubbing, or even in a cheep. A German expression likens a smacking kiss to a cow pulling its hoof from a swamp.

One proverb calls a beardless kiss "an egg without salt," and Danish philologist Christopher Nyrop, author of *The Kiss and Its History* (1901), felt women preferred the kiss of a bearded man. According to this authority, the most refined women of

Jutland say, "Kissing a fellow without a quid of tobacco and a beard is like kissing a clay wall." But there are others "who are not so particular in the choice of words, and these latter say straight out: 'Kissing one who neither smokes nor chews tobacco is like kissing a new-born calf on the rump.' "

Unexpected kisses can be potent, transforming potions. In Chekhov's "The Kiss," the love-starved Ryabovich wanders into a dark room where a lady awaiting her swain kisses him eagerly, and over the next few days, the act blooms in his mind into passion for a woman he cannot even visualize. In the pastoral romance *Daphnis and Chloe* (c. 200 A.D.), Chloe steals a kiss from Daphnis. He reacts as if stung, yet he shivers and his heart pounds. Chloe's face, which he'd noticed no more than a toadstool before, suddenly dazzles him and he blushes violently. "My breath's coming in gasps, my heart's jumping up and down, my soul's melting away — but all the same I want to kiss her again," he says. "Oh, what a strange disease — I don't even know what to call it."

The kiss has long emblemized the alchemy of love. It awakened Sleeping Beauty and makes frogs into princes. In a Scottish ballad, an evil stepmother changes an earl's daughter into a snake, but the hero Kempion kisses her three times and she blossoms forth in human form. This legend has more variants than DNA. In one, the stepmother changes the maiden into a lime tree, and she stands pinned to the ground for ten years until the king's son kisses its root. In Boiardo's *Orlando Innamorato* a beautiful damsel sits by a tomb in a castle. She urges the baron Brandimart to open it and kiss whatever he finds inside. He lifts the stone and out springs a serpent with brilliant eyes and fangs. Quaking with fear, he gives it an icy kiss, and it becomes a golden-haired fairy who gratefully enchants his armor and horse.

Sir John Mandeville tells of Hippocrates' daughter on the isle of Lango. The goddess Diana turns her into a dragon, and she can become a fair damsel again if a courageous knight will kiss her on the mouth. At least two knights vow to do so, but both turn tail when they see her hideous face. She throws one off a sea cliff, and as the other flees, she bursts into a terrible wail.

The philosopher Favorinus of Arles said, "At what else does that touching of lips aim but at a junction of souls?" He might have asked kissing bandit Morganna Roberts, who liked to dash out onto the field and kiss baseball players. On May 2, 1988, Morganna planted a smooch on Baltimore Orioles shortstop Cal Ripken Jr. near home plate. Police arrested her for trespassing and she spent the night in jail. It was "Fantastic Fan Night." José Moura, Brazil's notorious "serial kisser," achieved his greatest triumph when he evaded security to kiss the feet of Pope John Paul II in 1980. In 1991 he tried to kiss Martina Navratilova during a tennis match, but police hauled him away.

A kissing bandit of a different sort was twenty-two-year-old Tabetha Dougan. In August 1994 she met a seventy-four-year-old man in a bar, went home with him, and slipped him a sedative in a kiss. When the groggy man awoke next morning, he found she had stolen his Rolls-Royce, a pocket-watch collection worth $100,000, and $6,000 in rings and other jewelry.

Stolen kisses can be harmless, like Morganna's, or rapturous, like Chloe's. But they can also be odious. In nineteenth-century Naples, after a man kissed a certain woman against her will in the street, the courts forbade him to come within thirty miles of the site of the crime. Roman law called the offense *crimen osculationis*, the crime of kissing, but held it pertained only between people of equal rank when done unchastely. Kissing a nun escalated the penalty. In England in 1837, Thomas Saverland sued Caroline Newton, who had bitten off part of his nose when he tried to kiss her. The judge found for Miss Newton, declaring that "when a man kisses a woman against her will she is fully entitled to bite his nose, if she so pleases." In U.S. common law this crime is a battery, like any unwanted touching, and can also be sexual harassment.

Medieval literature abounds with examples of the kiss as an irresistible slippery slope. Men kiss women, then feel Jovian g-forces tugging them further. The troubadour Peire Vidal, for instance, said, "I entered her room and stole a kiss from her on the mouth and chin. That is all that I had. I am dead if she withholds the rest." It is the mentality of date rape.

Mistletoe is a celebrated catalyst for kisses, wanted or not. One caught under mistletoe must yield a kiss, a tradition some say arose from the plant's resemblance to human genitals. In 1855 Nathaniel Hawthorne complained of the custom in Liverpool: "The maids of the house did their utmost to entrap the gentlemen boarders, old and young, under these privileged places, and there to kiss them, after which they were expected to pay a shilling." In 1989 Moorhead State University in Minnesota banned mistletoe, on the grounds that it invited sexual harassment.

Kisses commonly express affection, as well as compassion, gratitude, reconciliation, wild joy, and deep sorrow. Family members and relatives commonly kiss, and the jealous Propertius charges Cynthia with inventing an army of kin to justify smooches from strangers.

We kiss at the start of long absences, as if to fill the future void with affection. The Old Testament abounds with farewell kisses, and when Paul left the Elders of Ephesus, they wept and showered him with kisses. In the ultimate departure kiss, people kiss the lips of a lover's corpse, as Propertius and Tibullus asked their women to do. Legend had it that a death-kiss kept the spirit in the body a bit longer. Ovid, exiled in Tomi, mourns that his wife won't be able to extend his sojourn on earth with her kisses. The ancients believed these kisses followed the decedent down to the underworld, comfort in the greatest void of all.

The kiss is often a gesture of greeting. Even chimpanzees embrace and kiss after a separation, one sign of the deep roots of kissing. People have welcomed the sun and moon with kisses, and individuals in many lands greet each other with a peck on each cheek. In ancient Rome greeting-kisses became such a nuisance that Tiberius (42 B.C.–37 A.D.), who detested flattery, issued an edict against them. It had little effect. "Every neighbor, every hairy-faced farmer, presses on you with a strongly scented kiss," Martial (c. 40–c. 102 A.D.) laments. "Here the weaver assails you, there the fuller and cobbler, who has just been kissing leather; here the owner of a filthy beard, and a one-eyed gentleman; there one with bleared eyes, and fellows whose mouths are defiled with all manner of abominations."

The custom persisted through the Middle Ages. Montaigne (1533–1592) complained of it. Why should a lady have to kiss any oaf with lackeys? he asks. And men fare no better, "for we have to kiss fifty ugly women to three pretty ones." Erasmus (1466?–1536) notes England was a rainfall of kisses. Even the pretty washer-girls at inns gave travelers beaming kisses as they departed.

This lip-largesse continued up through the seventeenth century. The marquesses of Molière (1622–1673) kiss each other freely, though in Le Misanthrope Alceste complains that Philinte kisses everyone and "when I ask you who it is, you scarcely know his name!" But the habit waned in the eighteenth century and has not returned, except in Hollywood.

Kisses of love have a vulnerability, like love itself, and the kiss long ago evolved into a sign of respect: I am vulnerable to you. The inferior offers lips, the superior a less sensitive surface.

The foot is a classic kiss receptor. In the Babylonian *Epic of Creation* chthonic deities called Anunnaki kiss the feet of Marduk. Mary Magdalen kissed the feet of Jesus, and the custom of kissing the feet or hands of saints continued for centuries. Caligula made subjects kiss his feet, and medieval vassals bussed the hands or feet of their lords. When the proud Rollo, a Norman chieftain, had to pay homage to Charles the Simple, he faced the noxious prospect of bowing to kiss the king's feet. Instead, he seized Charles's foot and lifted it to his lips, upending the king to the general amusement.

The ground is another well-worn kissing surface. In Mesopotamian myth, when Nergal visits Ereshkigal, queen of the Underworld, he kneels and kisses the ground before her. In ancient times, people kissed the footprints of their rulers, literally licked the dust. Isaiah promises that kings and queens "shall bow down to thee with their face before the earth and lick up the dust of thy feet." Nobles in *The Thousand and One Nights* kiss the floor before their sultans. When Raskolnikov confesses his murder to Sonya and asks her advice, she says, "Go now, this minute, stand in the crossroads, bow down, and kiss the earth you've defiled." He does, amid the hilarity of onlookers.

People have kissed altars, idols, and temples out of veneration. Cicero says reverent kisses had worn away the lips and beard of the statue of Hercules at Agrigentum. Kissing the cross is an especially holy act. People kiss images of the Virgin Mary and of saints, and kisses eroded much of the right foot of a statue of St. Peter in St. Peter's Square. A kiss on a saint's relic has cured ailments for centuries, and Pascal's niece overcame disease, we learn, by kissing one of the thorns of Christ's crown.

The intimacy of the kiss suits it for insults. The devil demanded kisses on his behind, and certain secret societies have required this act in initiation rites. In "The Miller's Tale," Alison pulls a similar trick on the mocked suitor Absolon. We have the slur "Kiss my ass" and the Germans say, more delicately, "He can kiss me where I have no nose." The Latin word for kiss is *osculum*, whose first syllable, *os*, denotes the mouth and the second two the rear end. In one story from ancient Rome, a man begged a kiss from a married woman, and she rejected him laughingly: "My first is for my husband, not for you. But you're right welcome to the other two."

A kiss is a promise. The kiss of love vows fidelity and hence caps marriage ceremonies. The kiss of respect implies fealty. A kiss could seal a feudal alliance. After dubbing a new knight, the master of ceremonies often kissed him. In *The Song of Roland,* Ganelon swears to betray Roland by kissing the hilt of the Saracen king's sword.

But the promise can be shallow. On July 7, 1792, with the members of the French Legislative Assembly quarreling and the Austrian and Prussian armies marching on Paris, Antoine Adrien Lamourette eloquently urged deputies to swear eternal brotherhood. In response, they ran into one another's arms and exchanged kisses of reconciliation. The feuding recommenced the next day, and Lamourette died on the guillotine two years later.

The promise can also be false, and a kiss an ill omen. A Judas kiss is one with treachery at its heart. As St. Ambrose wrote,

"A kiss conveys the force of love, and where there is no love, no faith, no affection, what sweetness can there be in kisses?" Ambrose called Judas one "who by a bestial conjunction of lips bestows a sentence of death rather than a covenant of love."

We kiss objects in superstition, hoping to wring good fortune from them. Gamblers kiss dice for luck and tourists lean down to kiss the Blarney Stone. To prevent a lightning strike, some rustics make three crosses before themselves and kiss the ground three times. To cure a toothache, they kiss a donkey on its mouth. If a book or a piece of bread falls on the floor, they kiss it when they pick it up. Mothers kiss their children's scrapes to "make it better" — perhaps less a superstition than a pacifier.

Alfred Eisenstaedt photographed one of the most famous public kisses in history: a sailor kissing a woman in Times Square, on the day World War II ended. It's a balletic movie-poster pose, in which the man swings the woman's torso almost parallel with the ground, leans deeply over, and kisses her. It suggests a passionate reunion, but Eisenstaedt says the sailor was drunk and weaving down Broadway kissing every woman he met. It was a kiss of giddy relief and celebration.

But public kisses are not always so tolerated, even between intimates.

Cato (234–149 B.C.) punished a senator named Manilius for kissing his own wife in public, in front of his daughter. Plutarch criticized Cato for this act, but added, "How disgusting it is in any case to kiss in the presence of third parties." Clement of Alexandria (150?–220? A.D.) advised married couples to avoid kissing before their servants. The inimitable Nyrop writes, "One evening at a large party I saw a young girl ostentatiously kiss on the mouth the gentleman to whom she was engaged. Cato would certainly turn in his grave if he knew that such immodest behavior was actually tolerated by people of refinement and position."

Even in the West today, public kissing can spur censure. In 1991 a Los Angeles condominium association warned a fifty-one-year-old resident against "kissing and doing bad things" in public. She sued, claiming she now had to endure "degrading remarks and questions from strangers." The group apologized, alleging mistaken identity.

Asian and some African peoples traditionally have avoided public mouth-kisses, viewing them as purely private and substituting bows, "sniff-kisses," and other gestures. In 1897 French anthropologist Paul d'Enjoy noted that the Chinese looked upon Western mouth-kissing with horror, as almost a cannibalistic act. The assiduous Malinowski found no evidence of mouth-to-mouth kissing in Trobriander foreplay, but plenty of mouth-rubbing, tongue-sucking, tongue-rubbing, lip-sucking, and lip-biting.

Sniff-kissing, or nose-rubbing, is common in many cultures, and signifies an intermingling of two people's breath or spirit, the equivalent to life itself. Maori greet strangers by touching noses softly, twice. It brings two faces within a tight circle and establishes a closer bond than a handshake. But it's far from a kiss.

A kiss transfers a spark of soul, even into a foot or earth. It's a touch of tenderness, an oath, a weld of identities, and for a Westerner, at least, it's hard to imagine the world without it.

The Lively Hinterland

In *Paysage de Baucis* (1966), René Magritte depicts himself as a mere nose, mouth, and eyes beneath a bowler hat. These features float in vacant air, without musculature, bones, or outline. He looks depersonalized, for it is a face sapped of singularity and expressive range.

Beyond the well-known facial landmarks lies the matrix, the smooth skin of the cheeks and forehead. It seems an empty quarter, yet its muscles signal constantly and its parts even reach symbolic import. The cheeks are blazons of jollity and passion, and illuminati attach mystic significance to the forehead, home of the "third eye."

The cheeks are the soft flesh of the face, dropping from the eyes to the nasolabial folds. Why have them? They house the oral cavity and hold food for chewing. Hence they are mammal specialties, absent from reptiles like iguanas, though they do appear in some dinosaurs.

The cheeks are also a locus of facial excitement. They can be shallow, even gaunt, but joy flushes them and laughter swells them, so the chubby-cheeked seem merrier. Like the nose, they redden with drink, and Frans Hals's *The Jolly Toper* (1630) may be the classic ruddy-cheeked image. And "cheek," like "face," can mean effrontery, impudence.

At the same time, blushes find a ready seat in the cheeks. To admit embarrassment, we say, "Are my cheeks red!" In one study, 68 percent of interviewees said they blushed mainly on the cheeks, compared to 26 percent for the entire face. Emily Dickinson wrote, "A cheek is always redder/ Just where the Hectic stings." Against our will, the cheeks can signal desire, confusion, guilt, self-consciousness.

The forehead seems the seat of intellect, no doubt because of the organ behind it. We associate a high forehead like that of behaviorist B. F. Skinner with intelligence, and a "highbrow" is a lofty, often pretentious mind. In Poe's "King Pest," six corpse-like sots sit around a table, each with one huge feature—

forehead, nose, ears, cheeks, mouth, eyes — and the giant fore-head belongs to the king. If asked to point to the "you" in one's head, most people indicate a place just above the bridge of their noses. In *Self-Portrait as a Tehuana, or Thinking about Diego* (1943), Frida Kahlo painted her wayward Diego Rivera in the center of her brow. The eyes show thought and the mouth articulates it, but the forehead is its symbol.

Civilizations all over the world have sensed its emblematic nature. On Ash Wednesday, Catholics place gray smudges above the bridge of the nose. Nigerians once greeted each other by rubbing brows, a charming gesture. Muslims often develop a permanent bruise on the forehead from decades of touching it to the earth in prayer. In the Book of Revelation, the mark of the beast adorns the brows of the depraved, and the name of God those of happy dwellers in the New Jerusalem.

Swift made the forehead a site of specific omen. In *Gulliver's Travels*, a few rare children in Luggnagg are born with a red spot above their left eyebrow, indicating they are struld-bruggs and will never die. The coin-sized spot grows over the years and changes color, to green at twelve, deep blue at twenty-five, and permanent black at forty-five.

But no culture treats the forehead more extravagantly than India. Its people daub the brow with *tilaka*, marks of dizzying variety. One scholar says the forehead became the host for *tilaka* because Hindus touched their brows to the ground before holy relics and lifted them to heaven in prayer.

Its purposes are legion. The *tilaka* can show sect. For instance, three vertical lines (for Brahma, Vishnu, Shiva) identify a member of the Ramanujacari sect, while a lone vermilion dot marks an extreme Shakta, and a design like a test tube holding a spot indicates a Vairagi. Some sects require the *tilaka*. A man without it, they say, is ignoble and anyone who happens to glance at him must look at the sun to purify himself. Other sects ban it, saying it leads to certain ruin.

The *tilaka* best known in the West is the scarlet dot women place above the bridge of the nose. Custom ordains this tiny disk, and researcher Priyabala Shah notes that many Indian women "do not know why they put a red mark on the forehead."

Tilaka can be potent charms. One holy book notes that the proper *tilaka* gives one "victory over kings, proud women, mad elephants, lions, tigers, serpents, giants." Another states that, with the right *tilaka*, "man becomes king and his enemies consider him like a lion or a tiger in front of a goat." *Tilaka* can also be beauty aids and elements of ritual.

Some Hindus have applied red arsenic as the *tilaka*, and others have used turmeric, saffron, sandalwood paste, betel leaf, and red lead. Early travelers among aboriginal tribes in India saw them sacrifice a human being and use his blood to dab a *tilaka* on a worshipper's brow. Orthodox families fear a forehead blank even momentarily, and tattoo the mark in place.

Like the *tilaka*, the notion of the third eye reached zenith in India. Some scholars believe it was flourishing there as early as the time of Buddha (6th cen. B.C.), who, tradition says, was born with a third eye in mid-forehead that took the shape of a hairy mole, called the *urna*.

The third eye normally refers to an organ that perceives inner essence, and it has an intriguing relationship with the recent quest for the nature of consciousness. Why do we have a sense of self? How does awareness of our inner workings benefit us? We could presumably make some good decisions without it, but social judgments would baffle us. Nicholas Humphrey, a psychologist at the University of Cambridge, suggests its function lies here. The self is an inner model of others. We sense our own reactions and project them onto others. We can thus predict their responses and act accordingly. The self, Humphrey says, is a cicerone to others' souls. He even calls this skill an "inner eye."

Of course such insight confers social power. In Hindu myth, the third eye is more: an extraordinary weapon. Turned outward, it emits a fiery beam that turns its target to ash and destroys the gods and all living beings at each periodic annihilation of the universe. The third eye also yields the aphrodisiac *madhu*, which flows forth like the Ganges.

What really lies behind the forehead? In fact, it is the frontal cortex, which not only helps control emotions like anger, but plays a key role in judgment. If wisdom is a matter of decision-making, then the third eye has hit the bull's-eye.

Within the Helix

In 1731 a Spanish sailor, defending his nation's trading rights in Cuba, boarded an English brig and sliced off the ear of Captain Robert Jenkins. This impertinence outraged the captain, though it didn't alter his appearance, since he wore a wig like most other sailing officers. But he kept the ear, pickled it, and talked constantly about it. Seven years later, when relations with Spain had grown tense, Jenkins displayed it indignantly to Parliament. Public ire soared, and the government reluctantly declared war on Spain: the War of Jenkins' Ear.

As ears go, Jenkins's enjoyed a remarkable trajectory. The ear is normally a humble facial feature, inconspicuous and often buried in hair. In Haruki Murakami's *A Wild Sheep Chase* (1989), the copywriter-narrator becomes fascinated by enormous photos of an ear in his office, "the quintessence, the paragon of ears." He feels the erotic pull of its whorls, and eventually seeks out and sleeps with the woman who bears them. But this is a whimsy, and the effect stems from its oddity. The ear is a facial outpost.

Hence big ears can strike us forcibly. The modest suddenly clamors for attention. Everyone's ears lengthen with age, but some stand out long before. The gangster who dominated Shanghai's drug trade in the twenties and thirties—and tutored a choleric thug named Chiang Kai-Shek—earned his sobriquet from his ears: Big-Eared Tu. The 45-foot Buddha in Cave 20 at Yungang, China, has epic ears that touch his shoulders. In the Mesopotamian *Epic of Creation*, godlike Marduk has four colossal ears (along with four eyes and a breath of fire).

Fish lack true ears. Rather, they have a lateral line system, which detects quivers in water. This apparatus can lie deep within, since fish are about as dense as water and vibrations travel right through them. But sound waves in air are too weak for the lateral line system, and when the first fish struggled onto land, they needed a way to sense them.

They developed both the eardrum and the hammer-anvil-stirrup system, which cleverly magnify air waves. (In a famously weird evolutionary move, the jawbones migrated upward to form these tiny structures.) The intensified signals reach the snail-shaped cochlea, where they cause the fluid inside it to vibrate. The cochlea has minute hairs sensitive to these pulses, and they send messages to the brain: sound. In a sense, we still hear underwater.

Land animals also sprouted a pinna, or ear flap. (We are so used to calling the ear flap the "ear" that "pinna" smacks of jargon, like calling the cheeks the "buccae." But the distinction can matter: Without pinnas we are lessened, but without ears we are deaf.) All primates possess pinnas, and land mammals without them are rare. Why have a pinna? Darwin deemed it useless. Cicero came closer. He thought it amplified sound, like the box of a violin, and it does, somewhat. But its main job is different.

The pinna helps locate sound. Its ridges and clefts bounce a few sound waves into the ear later than the rest, in a pattern that depends on their source. The brain then decodes it. Scientists have filled subjects' pinnas with wax, and found they perceive sound as coming from inside the skull, as with headphones. Some convolution of the pinna is essential, but extra amounts don't seem to improve performance.

We also locate sound through the fact that we have two ears, which create a kind of auditory parallax. Sound waves strike one ear slightly before the other, and the brain notes the difference.

Even so, we are fallible. We think movie voices come from the screen, and ventriloquists regularly fool us. The long-eared bat (*Plecotus auritus*) does much better. Its scooplike pinnas are longer than its body and enhance its echolocation, so this extraordinary creature can fly through scrub twigs at night and catch insects in rain. Like many animals, it can also move its pinnas to focus on sound. Among humans, the talent is rare, the stuff of classroom glory. We turn our heads instead.

In some animals, the pinna helps cool the body. The desert-dwelling jackrabbit has huge ears richly veined with blood vessels, which dissipate heat. The ears of elephants do the same. These great beasts are somewhat spherical, and hence must disperse more heat per square inch of skin. By flapping their fanlike ears, they speed the process. The snowshoe hare, on the other hand, has short ears, to minimize heat loss.

The ear has a special geography almost no one ever notices. The rim is called the *helix*, after a fancied resemblance to a coil, and it curves in over the pinna like a breaking wave. A second ridge abuts it halfway down: the *antihelix*. The antihelix swings up into a little plane and down into the *lobe*. While most of the ear is cartilage, the lobe is soft and fatty, just right for

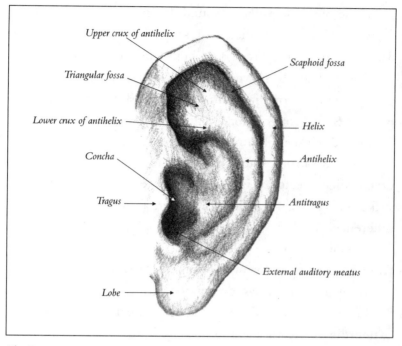

The Ear.

hanging ornament. The hollow near the ear canal is the *concha*, from the Latin for "shell."

And the little nub of flesh beside the ear is the *tragus*. What child hasn't fingered this curious flange and wondered what it does? In fact, it protects the ear canal. The name, Greek for "he-goat," stems from the hair on its inward side. It suggested a goat's beard to Rufus of Ephesus, a contemporary of Pliny and the first medical lexicographer, the man who christened the tragus as well as the helix and lobe.

Each person's ear patterns are unique — like his face, irises, fingerprints, handwriting, voiceprint, scent, and contours of facial heat emission. They thus offer the potential for identifying criminals. In one famous case, investigators compared photos of the ears of Anna Anderson with those of Anastasia, the youngest daughter of Czar Nicholas II, whom Anderson claimed to be. They found many similarities, but poor photo quality stymied a final identification, and they merely concluded that they couldn't rule Anderson out. DNA tests later revealed her imposture.

The best-known ear in literature probably belongs to Hamlet's father. Claudius killed him by pouring poison into it as he slept, a metaphor for pernicious advice, that specialty of Shakespeare's villains. And the best-known ear in history is not Jenkins's but Van Gogh's.* He sliced it off after an argument with Gauguin, wrapped it carefully in paper, and gave it to a prostitute in a nearby brothel, saying, "Here. In remembrance of me." Neither is as famous as Pinocchio's nose, and it shows how peripheral this feature is. The ear is a facial Connemara, nestled up against a wild sea of hair.

Berenice and Blackbeard

A treasure frames the upper face: the hair. Some 100,000 strands adorn the head and they give the face a silken backdrop. It is as if the face exists within a fine-grained lushness, beyond ken and faintly fragrant with desire.

A gleaming eye can stir and the teeth can dazzle, but light from hair is subtler. It has a strange, almost bewitching delicacy, and studio photographers use a special spot, the hair light, to emphasize it. Apuleius (c. 2nd cen. A.D.) rhapsodized about women's "beautiful scarf of love" and said that though a lady might deck herself with gold and gems, her striving was doomed without rich, glittery tresses. In John Updike's *Brazil*, Isabel has "luminous hair, full of many little lights like Rio at night seen from Sugar Loaf." Berenice's Hair, the only constellation named

---

*Evander Holyfield's may or may not eclipse it.

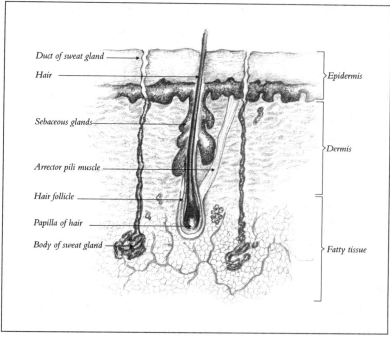

The Hair and Skin.

If hair is so useful and desirable, why does baldness occur? We don't know. According to one theory, it conveys age-related dominance. It exposes more cranial skin, so when the face flushes in anger, the display is larger and fiercer. Another theory suggests almost the opposite. As males age, they produce less testosterone and become more nurturing and grandfatherly, a trait which helps pass on genes. In the past, baldness signaled that such men were no longer the rivals of young warriors, and thus spared them physical attacks and made them more effective mentors.

Baldness is a tonsure from the genes, but humans have compelled similar restraints. In 1645, one year after conquering Beijing, the Manchu regent Dorgon ordered all Chinese men to shave their fore-scalps and grow a Manchu-style queue, on pain of death. Many Chinese treasured an elaborate coif as a sign of masculinity, and armed rebellions broke out. Dorgon prevailed, and the custom lasted till the twentieth century.

He had violated a widely accepted human right, even a Samsonish source of power. Longer hair lets us easily change our faces, and indeed hair is the most manipulable body feature. Its possibilities have fascinated people. Powhatan Indians of Virginia shaved the right half of their heads and let hair grow full-length on the left. Samurai cropped two deep coves into their hair. Mohawk Indians of course wore the "mohawk," a strip of hair from brow to nape which, according to a Dutch observer in 1644, "stands right on end like a cock's comb or hog's bristles." The Cherokee and Creeks often flaunted one as well, and some Creeks left an inch-wide cross on their pates.

Today we see initials, designs, and messages carved into the pelt, as well as beehives, brush cuts, ponytails, cornrows, bangs, pageboys, braids, pigtails, dreadlocks, green and orange spikes, a neverending sculptural variety. We put barrettes and combs in hair, and in Congreve's *The Way of the World* (1700) Millamant pins up her hair with letters — but only those in verse. Indeed, the hair feels like nature's gift to the imagination. We are born with our facial structure, but anyone can reshape the hair.

Beards and mustaches may be the most baffling facial features of all. Historically they have symbolized virility and dignity. Until a few centuries ago, men swore by their beards. Muhammad had a full beard, which he perfumed, and Muslims still take an oath "by the Prophet's beard." Some religions insist on whiskers in male devotees. Hasidic Jews take Leviticus 19:27 literally ("Ye shall not . . . mar the corners of thy beards"), and many Sikhs never cut body hair. For Muslim men, sporting a beard is *sunnat:* one earns credit for it, but suffers no penalty for abstaining. In at least one Muslim sect, a prayer leader must have a beard. In all these cases, U.S. courts have treated anti-beard policies as infringements on freedom of religion.

As a mark of dignity, the beard naturally invites affronts. When the Patriarch of Constantinople visited Louis XI (1423–1483) and angered him, the French monarch gripped his beard and led him around the room by it. A clutch of colorful phrases, mostly archaic, show the insult potential. "In spite of his beard" is "directly against his will." "To his beard" means to his face, brazenly. "To be in his beard" means to oppose with effrontery. "To put something against his beard" is "to taunt him with it." Drake's 1587 burning of the Spanish fleet at Cadiz was "a singeing of Philip's beard." And of course the modern verb "to beard" means to defy openly.

The abused can respond heartily. After John II and his entourage mocked the beards of Irish chieftains in 1185, they formed an alliance and defeated him in battle. When the Ammonites sheared David's ambassadors, he declared war.

Yet despite their apparent dignity, millions of beards vanish every day under the blade. Indeed, beards are the only part of the face we routinely excise.

Shaving is ancient and primitive tribes use seashells as razors. Folk wisdom to the contrary, severing hair does not accelerate its growth. Yet shaving can cause a medical problem for many blacks and some curly-haired whites. It's known as pseudofolliculitis barbae (PFB), or "beard bumps," and it resembles acne. The tiny beard hairs curl back and reenter the skin. Doctors call such hairs "bucket handles." The simplest cure is to grow a beard. The U.S. Marine Corps, which forbids beards, discharges men with this condition.

To shave or not to shave? Cultures often answer this question for individuals, and beards have ebbed and flowed erratically in fashion. In Nineveh men wore them curled and oiled, and often hardened them with perfumed gum and resin. Persians wove golden threads into them. In ancient Egypt nobles monopolized the beard, and it so symbolized power that Queen Matshrtpdont wore a false one, jeweled and gilded. The ancient Hebrews valued beards as marks of wisdom, and depicted Jehovah, Adam, Noah, and the prophets with lush beards. In the Middle Ages most male saints had whiskers. When Christians depict God, it's usually as an old man with a beard.

The ancient Greeks and their gods wore beards, and Diogenes mockingly asked the shaven, "What sex are you?" In Republican Rome, the smooth chin was likewise considered effeminate, but not in Imperial Rome. Beards helped bifurcate Christianity: Latin priests shaved, Eastern Orthodox priests grew matlike beards. Beards were out of fashion at the time of the Norman Conquest, though both sides wore long mustaches, but the Crusades temporarily revived them. Frederick Barbarossa (1122–1190) was Frederick of the Red Beard, from the Latin *barba*, beard.

By the early sixteenth century beards had vanished again. Henry VIII (1491–1547) revived the fashion in England. His jowls sported a bushy beard, and he required facial hair of his courtiers as well. By the time of Elizabeth most eminent men had followed suit, including Shakespeare, Cecil, Bacon, Raleigh, and Spenser. Special variants appeared—the Dutch, the Old, the Court, the Italian—and some marked a man's profession. A minister wore a Cathedral, for instance, and a soldier a Spade.

From the mid-seventeenth century through the mid-nineteenth, beards were rare in Europe. Voltaire, Newton,

Jefferson, Diderot, Washington — all were clean-shaven. Indeed, a furry chinstrap became a mark of eccentricity, or worse. The fanatic preacher Johann Edelmann (1698–1767), "The Notorious," wore a long beard, and when artist Liotard of Geneva grew a beard, it became as famous as his pastels and miniatures. The pirate Blackbeard (Edward Teach, 1680?–1718) had whiskers up to his eyes, we are told. He wove ribbons into this plumage, and according to one historian it "frightened America more than any comet that has appeared there in a long time."

Peter the Great (1672–1725) sheared a nation. When he took the throne, almost all Russian men wore beards, as if still trembling from the words of Ivan the Terrible (1530–1584): "To shave the beard is a sin that the blood of all the martyrs cannot cleanse. It is to deface the image of man created by God." But Peter felt beards showed Russian backwardness, and after his grand tour of Europe he ordered clean chins everywhere. Sometimes he shaved nobles himself, inviting them to a banquet and lathering them up. The shock to Russians was profound, but most complied, and eventually Peter offered resisters an out. For a fee, they could wear a humiliating bronze medallion around their necks that read: "Tax Paid." Beards returned only very slowly to the upper classes, and most Russians remained barefaced till the 1860s and 1870s.

In Western Europe, they began creeping back in the mid-nineteenth century, legitimized by writers like Dickens. The droopy sideburns called muttonchops or dundrearies appeared, along with the walrus mustache, which overhung the lips. European Romantics and Southern colonels in the Civil War sported goatees, and Chopin wore a beard on half his face.

By the *fin de siècle*, beards were in retreat. In *Trilby* (1895) the gauche Svengali wears a black, pointed beard, just as they were becoming passé. Every president from Grant through

Cleveland had facial hair; since then, only Teddy Roosevelt and Taft have. Beards returned in the sixties as banners of rebellion, outraging the shaven and helping polarize nations.

For a genetic feature, this is a remarkable history. Where is the evolutionary advantage in an item whole societies often erase? Beards are not even universal. They appear only on males, and the jaws of East Asian and Native American men are almost hairless. Why did beards appear at all?

It was probably not to warm us, as they occur in the tropics and not among the arctic Inuit. But they may keep us cooler. British archeologist A. M. W. Porter suggests that, since perspiration works mainly when it evaporates, beards may be sweat-catchers, useful to hunting men but not to women. Porter notes that recent hunter-gatherers like the Bushmen of the Kalahari could chase prey for hours, even days, so those men who could best keep their bodies cool would kill more game and win the nod from natural selection.

Beards first emerge after puberty and thus also indicate sexual maturity. They may send a stronger message. Psychologists Frank Muscarella and Michael Cunningham suggest they proclaim a man's aggressiveness, and thus aid his reproductive success. Since the signal is vaguely unsettling, some cultures suppress it. They can come to view the beard as vile and compelling, a fascinating beast-mark, as in cowboy melodramas of the fifties.

The two scientists propose that beards enlarge the apparent size of the jaw, triggering a response left over from the days when our teeth were weapons. They do bolster "weak" chins and recontour double ones. Yet they hit a slightly different note than large chins, more feral and mysterious. Beards may also work by suggesting the chin-jut, a threat among both humans and chimps.

Do beards attract women? The studies disagree, a fact that further snarls the problem. Hence these theorists suggest beards may boost a man's actual aggressiveness, perhaps by intimidating beardless men, and thus win women indirectly.

Smooth faces arose in monkeys to clarify facial signals. Did beards arise to obscure them? If beards make the lower face less readable, they presumably enhance deceit. In theory, bearded men could negotiate better and utter more credible political lies, and thereby garner more wealth, status, and wives. In turn, we may have evolved to regard them with suspicion and interest.

But ultimately the beard is an enigma, one more marvelous fact about ourselves that we simply don't understand.

II Panoply

Panoply

> It is the common wonder of all men, how
> among so many millions of faces, there should
> be none alike.
>
> SIR THOMAS BROWNE

2 Two   The Genes' Signature

I N 1553 Arnaud du Tilh was traveling through southern France
when two villagers hailed him. They thought him a resident of
nearby Artigat who had disappeared five years earlier. Du Tilh
corrected them and, intrigued by the error, quizzed them about
this doppelgänger, his habits, family, and neighbors. In 1556, du
Tilh presented himself in Artigat as the vanished man. He
seemed to recall old friends, and the man's wife accepted him in
her bed, perhaps realizing he was an improvement over the surly
youth who had deserted her.

Du Tilh posed as Martin Guerre for three years, and might
have continued if he hadn't sued Guerre's uncle to recover a
stretch of land. The suspicious uncle brought du Tilh to trial for
imposture, a serious crime. Du Tilh lost and appealed, and his gift
for improvisation might have saved him if the real Martin hadn't
suddenly arrived, cold with fury. The court expressed its awe and
regret to the prodigy, then had him executed. Soon after, a jurist
wrote a popular book about the case, and in 1982 the film *The
Return of Martin Guerre* resurrected it for a modern audience.

Du Tilh had performed a startling feat. Most imposters feign a profession, like surgeon-sheriff-monk-professor Ferdinand Waldo Demara, Jr., or a gender, like female pirate Anne Bonny. But du Tilh posed as a living face. He stepped into a real person's life, fooling his friends, kin, and possibly his wife, slipping past our multiple cues to identity. His exploit highlights the importance of recognizing faces. The Guerres were a respected family in town, and his performance made him, briefly, a man of substance.

The face is a signature in flesh and bone, and it remains our frontline against imposture. Faces adorn drivers' licenses, passports, credit cards, corporate ID cards — any document that might require proof of its link to us. Our singularity is our security, and our restraint.

It works because we live in a universe of faces, yet can recognize a single one instantly. We have special brain circuits for the task, and we're so adept at it that when we err in identifying criminals, juries usually believe us. Likewise, we can tell a male face from a female as easily as we blink, yet most of us have no idea how we do it, and scientists have only recently divined the key clues.

We take our uniqueness for granted, and if we need any proof of it, we need only look at faces that have doubled genetically and seem, to most people, the same.

The Living Double

Arnaud du Tilh actually resembled Martin Guerre only roughly, and he compensated by his wit. Rulers find better doubles. Pompey had two, says Pliny the Elder, and both could mimic his noble air. The encyclopedist adds that after Antiochus Soter (324–261 B.C.) was killed by his wife, she used an Antiochus lookalike to urge her own accession to the throne. Present-day assassin magnets like Saddam Hussein also send walking replicas out to public ribbon-cuttings.

But the best doubles are identical twins.

Identical twins are like mirages, Bali to the left and Bali to the right, conjurations to disorient us. Tweedledee and Tweedledum look indistinguishably petulant, and in *Lord of the Flies* (1954) William Golding links the two bodies of Samneric by one name and mind. The sense of mental interflow is understandable. Some 40 percent of twins develop a secret childhood language, opaque to outsiders.

Primitive people often held twins in awe and horror. According to explorer Mary Kingsley, residents of the Niger Delta regularly killed them, and sometimes their mothers as well. Aborigines, Inuit, and many African and Asian cultures also practiced geminicide, and in some corners of the earth it likely lingers.

Enigmas still hover about twins. For instance, we don't know why they occur. As biologist George Williams notes, twinning was maladaptive before people developed the technology to aid the survival of both. Even today, the global mortality of twins is about one in ten, compared to about one in a hundred for singletons.

Twins come in two kinds: fraternal and identical. Fraternal twins normally look quite distinct, like hairy Esau and smooth Jacob. The mother produces a brace of eggs and a separate sperm fertilizes each, so they are just two siblings in one pouch. But identical twins are forked existences. One sperm fertilizes one egg, which splits within the next 14 days. A single person doubles and both parts have the same DNA. They are clones.

Yet identical twins are never identical. Their faces, like their fingerprints, are always distinct, and parents and close friends can easily tell them apart after infancy. Some twins differ in eerie ways, such as sidedness. Their faces are mirror images of each other, right down to the hair whorls and teeth patterns, so each sees the other above the sink in the morning. In a few cases identical twins look no more alike than normal siblings. One can even have a cleft lip or muscular dystrophy and not the other. But if they possess the same genes, how is it possible?

The answer is that genes don't fully mold the face, and we don't even know how much they shape it. Differences in womb conditions can affect subtle processes like gene regulation and tissue feedback systems, and thus alter faces. Consider cleft palate. If one identical twin has it, the chance is 22 percent the other will too, as opposed to 4.6 percent among fraternal twins. It is both genetic and not genetic. A flawed gene apparently enables the defect, then an unknown environmental twist brings it forth. Altering womb chemistry in the right way suppresses cleft palate, and that's why women who take folic acid slash the risk of it in their babies.

More dramatic womb differences also send twins down separate paths. One may get more blood and food than the other. For instance, if one has a placenta feebly attached to the womb, it may become a much smaller version of the other. The junior twin may be lucky to survive at all. Ultrasound often reveals twins in utero, only to find one twin is gone later on, a phenomenon called "vanishing twins." Indeed, when doctors detect twins before ten weeks, single births ensue in up to 78 percent of cases. Sometimes the only external sign of the loss is vaginal bleeding.

Occasionally the genes alone make the difference. For instance, there are rare cases of "identical" twins of different sexes. The boy is born with X and Y chromosomes, and the girl with only one X. She is missing her twin's Y and so becomes a girl.

Identical twins can fool strangers to the point of chaos in plays like *The Comedy of Errors*, but friends and family neatly distinguish them. They approach copyhood, yet still differ from each other in myriad ways. Even in the closest genetic cases, the topography of each face is new.

"All Bubukles, and Whelks"

A cavalcade of faces passes before us in life, a stupendous stream from crib to coffin, and they are all basically similar: two eyes, a nose, and a mouth, in the form of a T. Yet we recognize a person's features swiftly, even after gaps of fifty years. We also adjust for changes in age, expression, background, and angle of vision. As computer scientists know, it's an astonishing skill.

We're deft at it. Lab research shows that people can recognize faces first seen two days earlier with 96 percent accuracy, and some studies have found little drop in score after a week and even months. In one experiment, identification accuracy declined only slowly for four months, but sharply thereafter.

The talent is especially distinctive because faces are within-class objects. They are instances of *human face*. We also excel at broad, basic categories, telling the difference between a bear and a moose, say, or a chair and a sofa. And we're good at recognizing items we own, like *my car* and *my hat*. But we struggle with other within-class objects, like specific bears, chairs, cars, and hats. Yet we can pick a single face out of the hundreds of thousands we've seen, and it doesn't even feel like work.

The ability kicks in early. Babies almost newborn recognize their mothers. In one experiment, researchers presented infants averaging 1.7 days old with their mother and with a woman who had just given birth and somewhat resembled her. They fixated on their mother's face 60 percent of the time.

Likewise, babies between 12 and 36 hours old make more sucking responses to a videotape of their mother's face than a
stranger's.

But we do not come into the world fully adept at face
recognition. The skill matures. For instance, youngsters under
eleven have trouble accommodating changes in age or expression. By adulthood, they handle them easily.

Yet several factors blur recognition. Faces seen in darkness
or barely glimpsed, like Harry Lime's as he darts through Vienna
in *The Third Man*, are harder to identify. A shifted pose or
expression can also challenge memory, at least of still photos.
One study found accuracy rates of 90 percent with a different
pose, 76 with a different expression, and 60.5 with both. Adding
a beard or wig cuts these scores further.

A sense of depth is important. We recognize a photo of
Tom Cruise better than a line drawing, partly because its shadows show the dips and rises in his face. We have a bedrock sense
of where this shade belongs, and if it strays, recognition tumbles.
In one study, people identified 95 percent of famous faces from
black-and-white photos, but only 55 percent from negatives.
Faces lit from below are also harder to recognize, because the
darkness settles in unfamiliar hollows.

We identify odd faces more quickly, easily, and confidently
than typical ones. Bardolph's spectacular visage, "all bubukles,
and whelks, and knobs, and flames o' fire," was an instant ID.
Jimmy Durante's nose became his insignia, as did Martha Raye's
mouth, and many other performers have stressed their facial
anomalies. Distinctive faces adhere to the mind, and it's a vig-
orous effect. Scientists have found it with exposures as short as a
second, and over testing intervals of four weeks.

Yet we don't recognize faces mainly by their parts. Rather,
we sense the array. Computer-image studies of the face show

we are exquisitely sensitive to the placement of features. In certain cases, a tiny shift in the mouth or eyes will lead us to perceive a completely different person.

It's not surprising, since the brain is geared to patterns. It can stumble with details and rules, but it identifies configuration brilliantly, dextrously, ceaselessly. It even plucks patterns out of noise, finding shapes in random letters on a page and conspiracies in tangled events. They all leap out at us whole, quite distinct from their parts. Hence we can tell a face is Oprah Winfrey's, for instance, without being able to say why. The brain is so good at recognizing these patterns that it doesn't leave an evidence trail for us.

What are the real parts of the face? In terms of recognition, they're not quite the organs we'd expect, like the eyes and nose. Instead, the face has pieces like tectonic plates, and scientists discovered them because of TV ratings.

In the late 1980s Arbitron wanted to invent a "people meter," a box that would sit atop the TV and discern the number of people watching and their attention level. It turned to the MIT Media Lab, where Alexander "Sandy" Pentland began working on the project. Arbitron later dropped out, says Pentland, because it worried how sponsors might react if they saw actual viewers, who might be doing homework, sleeping, or eating dinner in the next room.

But this research led to a device for recognizing faces. Pentland first compiled a face database, or facebase, of a few hundred people. He then sought to isolate the components of the face. He determined how much each face varied from the average, and his computer sorted out which deviations tended to occur together.

He discovered a mosaic of about 100 pieces, which he called "eigenfaces," from the German prefix *eigen* meaning "own" or "individual." An eigenface is an independent facial unit. Variations in one do not affect those in any other. Some are single swatches of territory. For instance, one lies on the upper lip and another on the forehead. Others are archipelagos, so one eigenface, for instance, includes parts of the underjaw, lower nose, and eye region. Eigenfaces are building blocks — they just aren't always coterminous ones.

The camera can detect about 100 different levels for each eigenface. Pentland's research thus reveals 100 facial provinces with at least 100 varieties each. Assuming all are genetically feasible, that makes at least 10^{200} possible human faces. It's a preternatural sum, far beyond any human sense of infinity.*

Using these novel categories, Pentland found he could describe anyone in his facebase. The whole discovery surprised him. "Face recognition was supposed to be very hard and complicated," he says, "yet here was a very simple method that worked very well."

It identifies people with a basic match-up technique. The software first analyzes a subject's eigenfaces, and in most cases it needs to assess only about 20. The computer then compares them with those in its facebase, a fairly easy matter.

---

*10^{200} is 10 multiplied by itself 200 times, that is, 1 with 200 zeroes after it. In comparison, the human brain stores about 10^{18} bits of information, the universe has some 10^{87} subatomic particles, and chess allows around 10^{120} possible games. Hence for every chess game there are 10^{80} or 100,000,000,000,000,000,000,000,-000,000,000,000,000,000,000,000,000,000,000,000,000,000,000,000,000 different faces.

And a commercial one. "There is a huge market already underway to make sure ID cards are valid—fit the person carrying them—and that people don't have more than one," he notes. Fourteen states and four nations are using his software to verify drivers' licenses and other ID, he says. In South Africa a bank is replacing cash cards with facial ID for distributing retirement benefits. And the U.S. Army deems his program the most accurate facespotter in the world and is using it to screen entrants into nuclear weapons sites. Pentland believes he can improve it to the point where it sees through heavy disguises, by focusing on the bone structure near the eyes, the hardest part of the face to alter.

Beyond checking ID, face recognition devices may forestall crime. In crowded shopping malls, for instance, they could scan for people with criminal records, especially shoplifters. The instruments could also identify loiterers near terrorist goals like the White House, a help since bombers usually case their targets several times before striking.

In the long run, Pentland says, "Face recognition, and more generally interpreting people's faces, is a critical part of making machines, rooms, and cars be human-centric rather than technology-centric." Machines that recognize their owners and respond to a smile, for instance, will change the whole feel of daily life.

The Neural Face

The Fifth Marquess of Salisbury was notoriously bad at recognizing faces, even of close acquaintances. Once, standing behind the throne in court, he noticed a man smiling at him. He whispered to a neighbor, "Who is my young friend?" The neighbor replied, "Your eldest son."

It's possible he couldn't recognize faces at all. A rare handful of people share this condition and, though otherwise normal,
they wander in a sea of unknown countenances. This remarkable oblivion is called prosopagnosia (from the Greek *prosopon*,
"face," and *agnosia*, "lack of knowledge"), and it reveals much
about how the brain handles faces.

Prosopagnosics lose the facial pattern. They often don't
recognize friends or family members, or even themselves. The
face in the looking glass is commonly a stranger's, and one victim apologized after bumping into a mirror. To identify others,
these people rely on voice or other cues, and occasionally an
odd facial feature like a mole or long nose.*

Their lethe can transcend the face. Prosopagnosics often
have trouble distinguishing members of a class, like different
kinds of animals, flowers, and foods. But this effect is not universal, and in its rare, pure form, the ailment wipes out only
faces. One prosopagnosic did just as well as normals at telling
earlier-seen spectacles from new ones, a job most people find
harder than recognizing faces. Another, who became a farmer
after his illness, grew adroit at recognizing the faces of sheep, a
task that perplexes ordinary people. Whatever his strategies with
sheep faces, he couldn't apply them to humans.

Prosopagnosia stems from brain damage. It can arise suddenly. One man was speaking to his physical therapist when he
blurted out, "But Miss, what is happening to me? I can't recognize you anymore." He had had a stroke and become prosopagnosic even as he talked.

---

can recognize parts of the face, and prosopagnosia does not seem to affect
ng. One victim who could not identify faces, expressions, or even gender
sceptible to the McGurk illusion.

The brain has special areas devoted to the face. Observers discovered a face cortex in monkey brains in 1972. It straddles the superior temporal sulcus, a large groove near the top of the temporal lobe, the "rabbit's haunch" in a side view of the brain. Cells here react to a range of faces, including plastic models and photos of both monkey and human countenances. Yet they respond scarcely at all to most non-face stimuli — bars, edges, gratings, textures, brushes, snakes, food, spiders. They ignore pictures of faces with scrambled features, and even react poorly to line drawings. And though few react to isolated facial features, they do respond to a whole face missing, say, a nose.

Intriguingly, this cortex performs not just one job, but at least three. It interprets identity, expression, and eye/head orientation — the who, how, and where of faces. In monkeys, the superior temporal sulcus forms a kind of divide. The cells for identity tend to lie below it, those for expression and orientation above. Cut out the upper area and monkeys can still identify a face, but they can't tell where it is looking.

The orientation system divines social attention and reacts to head direction, since it suggests intent. Some neurons fire more to a full face, others to a three-quarter view, still others to a profile. In fact, every angle of the head seems to spur a distinctive firing. These cells also register gaze. Some react more to eye contact, others to a downcast look. Hence our sensitivity to the aim of eyes. It's wired into us, and prosopagnosics often don't know which of two faces is staring at them.

In humans, the entire apparatus is more complex. First, the face cortex has migrated from the superior temporal sulcus down to the very bottom of the brain, to the fusiform and inferior temporal gyri. A scattering of cell clumps there handle facial tasks. But they do not work alone, and PET scans show heightened activity across much of the brain during face recognition.

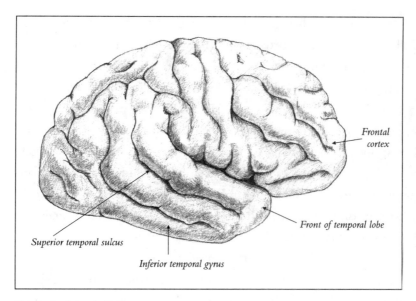

Frontal cortex

Front of temporal lobe

Superior temporal sulcus

Inferior temporal gyrus

The Brain (lateral view).

Some of these neurons perform other jobs as well, but the over-all number engaged suggests the complexity of this near-instant and effortless feat.

Consider: We recognize some faces, like the dimly famil-iar character actor in an old western, without recalling anything else about them. The face has simply welled up from anonymity to perplex us. This stark déjà vu is a first step, but true identity requires more. Normally, the brain also summons up a name, a dossier, and an emotional response.

The name circuit remains unclear. We may store names in the left middle temporal lobe. In one study, a patient with dam-age in that region could recognize faces and objects, but not name them. This subroutine lies somewhat apart from the rest of face recognition, since it provides a label rather than content. Hence a person's name can sometimes slip our mind, indepen-dent of all other knowledge about her.

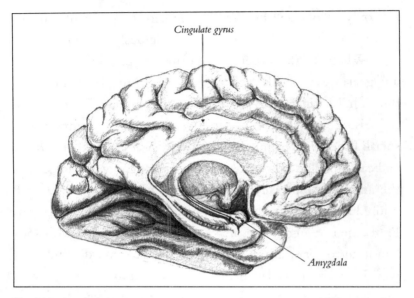

The Brain (medial view).

We may call up the dossier from the front of the temporal lobe. This area seems to store biographical lore on people and also comes into play in non-facial tasks. A patient named P. V., who had damage here, could recognize objects at the basic category level—that's a tree, a house, a person—but could not identify specific faces, nor specific monuments like the Eiffel Tower and Arc de Triomphe.

The most critical ID response is the friend-or-foe reaction, which can save our lives. Near the start of this century the French physician Edouard Claparede treated a patient who seemed unable to form new memories. Claparede had to introduce himself anew each time he walked into her room. One day he hid a tack in his palm and pricked her when he shook hands with her. The next time he entered, she refused to take his hand. Yet she still said she didn't recognize him and though she persisted in her refusal, she couldn't explain it. At some level she not only recognized his face, but linked it to a painful handshake. We can fear a face we don't consciously recognize.

Whatever the other damage in this patient's brain, her amygdala was probably working. The oval amygdala ("ah-MIG-duh-luh," Latin for "almond") is a busy nexus of brain signals that also contains neurons sensitive to faces. Its prime role is social: It helps determine feeling toward faces and other stimuli, and can quickly sound an alarm. PET scans show the left side of the amygdala especially reacts to fear and anger. People with injured amygdalas are typically cheerful, even blithe to danger. Frightened or angry faces make no impact, nor do words spoken in terror or wrath. These individuals are hazard-blind.

Does the amygdala handle other emotions? D. R., a fifty-one-year-old woman with a damaged one, was unable to recognize blends of expression, and researchers have suggested the amygdala helps discern mixtures of, say, happiness and surprise. Indeed, some scientists think it may be a clearinghouse for all social signals.

Yet it is plainly not essential for identifying them. In one 1997 study, for instance, a woman who'd had the organ removed could still recognize sounds of surprise, happiness, disgust, and sadness. In another using PET scans, the cingulate gyrus showed heightened activity when subjects were distinguishing happy, sad, and neutral faces. In yet another, the cortex responded to faces of disgust and the amygdala didn't. Moreover, parts of the right hemisphere may process fear independently of the amygdala. Nothing about the brain is simple.

Near the end of *Moby-Dick* (1851), Ahab stares at his shadow in the ocean, watching it lose shape in the depths. The face itself is like his shadow and its meanings can recede into darkness. When expressions need fine discrimination, like the face of La Gioconda and the smile of O. J. Simpson, they go to the frontal cortex, where we make most of our conscious decisions. Even there, an expression may lie just out of grasp, and the face can feel delphic.

Though the main face recognition system collapses in prosopagnosics, it works splendidly for most of us, most of the time. But now and then it misfires, and the upshot can be serious.

The Crime of Laszlo Virag

One day in 1969 a man was opening parking meters in downtown Bristol, England. A passerby asked what he was doing, and he replied he was collecting revenue for the city. The pedestrian grew suspicious and trailed him to his car, where the self-declared meter man drew a gun. The pedestrian fled and contacted police.

Patrolmen soon spotted the thief's car on the freeway and chased it. Twice he halted and brandished a pistol at them before resuming flight. Finally he abandoned the car and raced across a field on foot. The police dashed after him and he shot back, wounding one in the arm. He circled around to the freeway again, hijacked a car, and ordered the driver to take him to Bath. He had escaped.

But people had seen his face. Investigators soon learned of a similar crime in Liverpool. The criminal seemed to speak with an accent, and police knew that a number of Hungarians had committed meter thefts. So they showed witnesses mug shots of these suspects. Four fingered a man named Laszlo Virag. Officers brought Virag in and paraded him before seventeen witnesses, six of them police. Four police and three civilians named Virag as the culprit.

At trial their testimony formed the whole case against him. Police found no other link between Virag and the crime, yet he lacked an alibi and led a wastrel's life, subsisting on gambling and social security payments. The witnesses were very confident, and one officer said, "His face is imprinted on my brain." In July 1969, the jury convicted Virag and he began serving a ten-year sentence.

Two years later, police broke into the house of one Georges Payen and found the gun used in the Bristol chase and other evidence tying Payen to the meter thefts. They arrested him, and the government later pardoned Virag. One policeman later said Virag and Payen were "in no way similar in appearance and it is therefore difficult to believe that all these witnesses could have been mistaken." Even so, they proved reluctant to recant.

How could such a mix-up have occurred? No evidence indicates that the police intentionally biased the witnesses before the lineup. Rather, the case raises questions about eyewitness identification itself. How accurate is it, and how does it go awry?

Eyewitness error is a serious problem in criminal law. It is by far the leading cause of false convictions. In one recent survey of 205 miscarriages of justice, it tainted 52 percent of cases, with perjury a distant second at 11 percent. One scholar conservatively estimates 7,500 cases of wrongful felony conviction in the United States each year, of which eyewitnesses may cause 4,500.

It can happen to anyone. Australian psychologist Donald Thompson became interested in the phenomenon after officers arrested him and put him in a lineup, where a woman identified him as her rapist. Fortunately, he had an alibi: He'd been on TV with a police commissioner at the time of the crime. But if a jury ever falsely convicts you, chances are a witness has pointed an awful finger at you in court.

And that witness may be utterly confident. Juries tend to believe confident eyewitnesses, and so do judges. The leading U.S. Supreme Court case on admissibility of their testimony,

Neil v. Biggers (1972), specifies five relevant factors, and one is the certainty of the witness. Yet it is a bad plank to set one's keg on. Study after study has shown that the confident witness is only slightly more apt to be right than the hesitant one. What accounts for this well-documented finding? The answer is unclear, but plainly some factors boost certainty and not accuracy, and vice versa. For instance, a witness might grow more confident as he learns facts that seem to confirm his opinion.

Four of the seven witnesses against Virag were police. One might expect professional law enforcers to recognize criminals better than ordinary citizens. The evidence, so far slender, suggests they actually do, but only if they see the offender's face long enough. With brief exposure, as in this case, they perform no better than anyone else.

Payen wielded a gun in his crime. The presence of a weapon diminishes face recognition. A handgun attracts the eye, and even in videotape experiments it makes us neglect the face of the person brandishing it. In one such study, 46 percent of subjects correctly identified a robber without a gun, but only 26 percent recognized the same robber with one.

The witnesses initially picked Virag from an album of mug shots. The hazard in mug shots lies in showing people too many. Even 100 to 150 photos can reduce accuracy, and after 600 it falls markedly. A parade of faces passes before the witness's eyes, and they seem to overlay and alter memory of the target. Programmers are now developing software to shorten the search, mainly by showing witnesses only faces that resemble their descriptions.

A number of factors have surprisingly little impact on witness accuracy. Intelligence has scant relation to it. Richness of description is irrelevant. One witness can provide Vermeerian detail, another Cézannian daubs: both have the same identification rates. What about consistency? Police and lawyers typically question witnesses at the crime scene, in depositions, and in trial, and facial descriptions can change over this period. Such inconsistencies bear little relation to accuracy — an unnerving fact, since attorneys have long used them to discredit witnesses in court.

On the other hand, two fairly obvious factors do correlate with good identifications. One is freshness of memory. The more recent the viewing, the greater the accuracy, so the witness who identifies a suspect soon after the crime will do better. The other is exposure. The longer the eyewitness sees the perpetrator during the crime, the greater her accuracy. Numerous studies have shown that witnesses who see culprits only briefly, like most clerks in convenience stores, have high rates of false identification.

Many people and courts believe excitement jostles facial memory, playing hob with identification. It may. Yet studies on the impact of violence or eyewitness stress have yielded contradictory findings, possibly because the effect is complex. The question is hard to test empirically, since it is clearly unethical to put experimental volunteers in situations where they truly fear violence.

After a crime, police often ask witnesses to help reconstruct the miscreant's face, to prepare a sketch for other law enforcers or the media. The task is one of face recall, not face recognition, and it's a separate process entirely. The witness no longer sits waiting for a match, but must summon an image from memory and articulate it. It's the difference between recognizing the Golden Pavilion of Kyoto, say, and describing it so another person can identify it.

Traditionally, witnesses talk to a police artist, who draws a sketch and alters it as the witness dictates. But in the 1970s two new techniques appeared that seemed scientific and revolutionary: Identikit and Photofit. Both carve the face into parts and offer a raft of samples for each. Nose options, for instance, might include aquiline, pug, English, and broadly flat. The witness selects the suspect's nose and plugs it in. Identikit first used line drawings and later photos, and works by placing a series of transparencies atop each other. Photofit apportions the face into five segments, into which the witness slots a photo. Jacques Penry, the inventor of Photofit, stated its philosophy: "Because each facial part is the sum of its individual details and the whole face is the sum of its sections, the total assessment of it requires a careful visual addition."

Despite the early glow about these methods, sketch artists still surpass them, and even the drawings of witnesses themselves almost equal results from the kits. Their doom is Penry's "careful visual addition," which assumes that witnesses can actually identify a suspect's eyes, nose, chin, or mouth. They usually can't. For instance, in one experiment, 32 subjects tried to duplicate an image previously made from the five Photofit parts. Even with the living model before them, nobody succeeded, and the chin and nose proved especially vexing. With the model absent, they were correct in only 35 of 160 piece attempts. Their brains were sensing pattern, not part.

Intriguingly, this research has identified a key trio of factors which we recall more readily than any others. The first two are age and face shape. Both are matters of pattern and hence bedevil ID kits. The third is hair. In fact, hair is the easiest facial feature to describe to a sketch artist, a notable irony since we alter it so regularly. This finding means the astute bank robber will wear a cap. In one experiment, subjects viewed hatless men in a lineup. In half the cases, the men had gone bareheaded during the robbery; in the other half, they'd worn a knit pullover cap that covered the hairline. Witnesses identified 45 percent of the hatless, but only 27 percent of the capped thieves.

Today, computer programs like Mac-A-Mug and EFIT exist to help face recall. They allow much more flexibility in creating a face. Careful evaluations are so far lacking, but the same problems likely afflict them. Indeed, one scholar states that the average witness may find it impossible to re-create a suspect's face well enough to permit identification.

Crime tests the frontier of facial recognition. It shows where our capacities break down, and it also shows how readily we deceive ourselves about our own and others' abilities. No one knows how many innocent people remain in jail because of false identifications, nor how soon our cautious judicial system will acknowledge the recent findings on faces.

Nelson Mandela's Eyelids

Over the last centuries an inadvertent historical experiment occurred on tiny St. Vincent in the Antilles. First populated by Arawak and Carib Indians, the island became a refuge for escaped black slaves in the 1500s. The races interbred, and a team from the University of Kansas found that current residents had 43 percent Native American genes, 41 percent African, and 16 percent European. But these "Black Caribs" look very African. Their genes have predominated at the surface, which suggests their role in the tropics.

Race is in the face. We first notice it there — and notice it and notice it and notice it. The importance societies attach to race is arresting and faintly ridiculous, akin to believing TV characters are real people. Race is skin-deep and genetically trivial.

Indeed, until recently we were all black. The best current evidence indicates *Homo sapiens* left Africa around 100,000 years ago and settled Australia perhaps 60,000 years ago, though this date keeps sliding back. In Eurasia, "white" and "Oriental" forked around 40,000 years ago.

Why was African skin black, and why did emigrant skin change color? As Immanuel Kant suggested as early as 1775, race is mainly a response to latitude. It reduces death from exposure to too much or too little sunlight.

Black skin shields against too much. It stems from melanocytes, cells which form a layer at the base of the epidermis. They make a brownish–black sunblock called melanin, which protects the body from ultraviolet rays. These rays cause melanoma, a fast-growing cancer that can kill adults in their twenties and thirties. And even a nonlethal skin tumor would have invited a swarm of infections in prehistoric people. So in lands of high sun, the dark-skinned passed on more of their coding.

The surfer virtues of black skin are clear. Blacks can swim and sunbathe in much greater safety than whites. In Australia, aborigines rarely suffer melanoma and whites often do, despite their bottled sunscreen and bush hats. Indeed, the melanoma death rate in Cairns is the highest in the world. In the United States each year, whites average 4.1 to 4.4 cases per 100,000 people, blacks 0.6 to 0.7. Asians get less skin cancer than whites and more than blacks.

Melanin is a prize in many ways. It reduces sunburn and wrinkling. Rosacea, which requires sun damage, afflicts blacks less often and some have called it "the curse of the Celts." Melanin may keep ultraviolet rays from breaking down nutrients like folic acid, riboflavin, and vitamin E. It also prevents the body from making too much vitamin D, which causes calcium deposits and kidney stones.

Sunlight turns on melanocytes, so they spew forth melanin and darken the skin. Black babies are born pale and do not turn black until exposed to the sun. Freckles too appear where photons hit the skin, mainly the face, and grow lighter in winter. A person locked in perpetual gloom would have no freckles. One who always wore a bathing suit might have freckles on her thighs.

Sun brings out a tan, the pool lounger's treasure, which soon fades without the ultraviolet tattoo. As Ishmael strolls New Bedford in *Moby-Dick*, he uses face color to guess how long sailors have been ashore: "sun-toasted pear" means three days landed from an Indian voyage, "a touch of satin wood" is a week or more, a slightly bleached "tropic tawn" suggests several weeks.

A tan is armor against sunburn. A deeply tanned person, a "sun-toasted pear," may attain a sun protection factor (SPF) of 5, while a chocolate-black African has one of 6 to 7. An average T-shirt confers a 15 when dry, a 5 when wet. What do SPF numbers mean? The scale originated to rank sunscreens and measures how much longer it takes to get the first hint of sunburn with the product than without. A person who burns in 10 minutes can extend the grace period to 50 by developing a deep tan (10 x 5) and 150 minutes by donning a T-shirt (10 × 15).

Yet physicians view the tan as an injury response, like a scar. And unlike naturally dark skin with the same SPF, it boosts the chance of cancer. A black's skin comes free, but a tan has a price: continual ultraviolet exposure and its harm to the DNA. And since a tan delays sunburn, it can bestow a false sense of safety. In our well-tanned Western society, melanoma has increased 812.5 percent over the last 50 years, an amazing upswing which scientists credit to increased sunbathing and the thinner ozone layer. Because of spotty enforcement of the 1987 Montreal Protocol, the epidemic continues to grow. Scientists predict cases of skin cancer will double in the United States and Europe by 2100.

All races have about the same number of melanocytes. But in blacks they pour forth more melanin, and in greater variety.

When our species arose in Africa, the melanocytes likely all secreted sunblock at the same rate. As people spread north, the rate slowed, but the telltale cells remained.

In higher latitudes, the ultraviolet rays weaken and a new hazard appears: too little sunlight. The sun's rays cause steroids beneath the skin to churn out vitamin D, which the body needs to metabolize calcium. Youngsters with a dearth of this vitamin become calcium starved, and their bones soften and bend. We call this disease rickets. In cooler climes, the sun hangs low in the sky, the body makes less vitamin D, and rickets grows more prevalent. Indeed, human remains in Sweden from 15,000 to 10,000 years ago show bones and teeth deficient in calcium. Since melanin blocks sunlight, it cuts vitamin D levels even further. Pale skin arose to let the photons in. Northerners with little melanin got rickets less often and spread more genes.

Blond hair and blue eyes arose from this white-out. Melanin also colors the hair. Every strand begins tulip-like in a follicle bulb beneath the skin. There, as hair cells split and multiply, they absorb melanin from nearby melanocytes. The amount they take in colors the hair dark or light. Indeed, variations in two kinds of melanin account for almost all mammalian browns, reds, golds, and blacks in the hair and fur. The Arctic fox produces two coats of fur a year and in the winter does not secrete melanin, so its pelt is white.

Blacks' hair is usually kinky and densely matted. If a hair shaft is circular, as in apes, it hangs down straight, but if flattened, it twists into a spiral or a more exotic shape. The tightly wound hair of an Afro rises, and both cools the head and protects it from ultraviolet rays.

Blue eyes lack melanin. Instead, they have a scattering of white flecks from which high-energy light — blue — bounces off more vigorously. A similar phenomenon creates blue sky, and irises like those of Paul Newman and "welkin-eyed" Billy

Budd match the heavens. If a layer of melanin covers the iris, the eyes look gray, hazel, brown, or dark brown, depending on its thickness.

Over 80 percent of people who live around the Baltic Sea have blue eyes, and the mutation apparently began there. Yet there are enclaves of blue-eyed individuals in Africa and among American Indians, suggesting the trait has arisen elsewhere and not flourished. Why did it spread in the Baltic? Anthropologist Jonathan Kingdon suggests that the lighter hue of the blue iris makes widened pupils more visible, so they could better signal aggression or desire. The trait may also have spread through sexual selection, an arbitrary preference for blue in mates.

Asians in general have flatter and more delicate faces, as well as flaring cheeks, smaller noses, and straight, tourmaline-black hair. But to other races, their most striking feature is the eyelid. Whites and blacks have a fold in the eyelid, dividing it in half, so the lower part slides up under the upper. Asians have more fat and so the lid is often flat and continuous, though many do have this tuck.

Asians also possess the epicanthic fold, a droop of flesh over the inner or both corners of the eye. One old theory holds that it acts as an awning, blocking excess sun, though this notion dubiously postulates that Asians needed it because they lived near snowfields. But the Khoisan of southern Africa have this fold as well, and anyone can see it in Nelson Mandela, a part Khoisan. One scientist has suggested it is genetically linked to concave noses. It is a tagalong, with no intrinsic purpose.

Noses vary widely among the races, and as early as 1923, physical anthropologists A. Thomson and L. H. D. Buxton linked their shape to climate. In very cold or dry regions, noses tend to be longer and narrower. In hot, damp climes like the tropics, noses are short and flat. The anthropologists suggested that long, bladelike noses better warm and humidify air on its way to the lungs, and scientists today generally concur.

Is there such a thing as a Jewish nose? To most people, this term means a convex, hooked nose. In one 1952 study of New York Jews, involving 2,336 men, researchers found 57 percent had flat noses, 14 percent concave, 6.4 percent flared, and 22.3 percent, or about a fifth, convex. A study in Poland yielded similar results. In other words, the "Jewish nose" is actually uncommon among Jews.

People are better at identifying faces of their own race, a phenomenon called the "own-race bias." It is a fairly small effect, but all races demonstrate it and scientists have replicated it many times. Attempts to train people to overcome it have generally foundered.

What causes it? Studies suggest it is not prejudice. Lack of contact comes closer. Individuals who have seen few other-race faces do worst at recognizing them. But exposure is not the whole answer. In one interesting study in Zimbabwe, black students at a Harare school who had daily contact with whites recognized white faces much better than did low-contact blacks. Yet whites at this school, who had daily contact with blacks, gained almost no benefit. Perhaps the difference lay in necessity. The authorities in the school were all white, so blacks *had* to know how to identify them. Whites had no similar motive. Intriguingly, the blacks who recognized white faces well did somewhat worse with black faces.

In another study, Ruth Dixon tested British whites who were participating in an eight-week program in Ghana or Tanzania, both before they left Britain and in Africa before they returned. At the end of their African stay, they were significantly better at recognizing black faces, and slightly (though not significantly) worse at recognizing white faces. Familiarity, and a simple prod, seem to straighten the bias.

Race is not entirely skin-deep, in fact. Races have also evolved internal means to protect themselves from climate extremes. The U.S. military has found that soldiers of African extraction get frostbite more easily than those of European or Asian. On the other hand, blacks resist heatstroke better than whites, since their internal temperature remains more stable in spite of massive sweating. Inuit have more fat under their skin and their hands can do skilled work at temperatures when those of whites and blacks grow numb.

In some places, peoples have also developed defenses against local parasites, like the malaria-fighting sickle cell gene in Africa, or adjusted to special diets, as with the high rates of lactose absorption among milk-dependent pastoralists like the Tussi of Congo.

Indeed, skin color is a misleading guide to genetic similarity. Both Africans and Australian aborigines are black, yet they share the fewest genes of any peoples on the planet. Aborigines have black skin because it saves lives in the tropics. But whites are closer cousins to Africans.

And Africans differ among each other genetically far more than other races do. After the ancestors of aborigines, whites, and Asians left Africa, the Africans' DNA continued to diversify. Hence, if one categorizes humanity along strict genetic lines, there are many races within black Africa itself.

Even so, interracial differences are slim. In 1980, molecular biologist B. D. Latter examined how proteins vary both within and across race. He discovered that 84 percent of the variation came from within the race. In other words, we are genetically much more similar to many individuals in other races than in our own. Douglas Wallace, a Stanford geneticist, notes that all human races are genetically closer than two subspecies of African gorilla living a few hundred miles apart.

So where does this all leave race?

Some scientists suggest it has no meaning. This notion seems a healthy counter to its significance in the world, yet it is not quite accurate. Biologically, race is real, though it involves superficial traits and very fuzzy categories. But race more often lives its garish, strutting life as a social construct, as a tool humans use to define themselves and others. People read essence into unchanging surface and see a mirage of mind where there is only disease prevention.

The Facial Map of Sex

Chinese opera singer Shi Peipu posed as a woman in life and in 1964 began an affair with an unwitting twenty-year-old named Baernard Boursicot, a bookkeeper at the French embassy in Beijing. Their infrequent sexual relations always occurred in haste and darkness, beneath a blanket, with Shi Peipu hiding his genitals in his hands. In August 1965 the singer declared he was pregnant. This child miscarried, Shi Peipu said, but by that December another youngster was on the way, and while Boursicot was in the Amazon, he heard about the birth. Boursicot did not see the child until 1973, and in 1982 he finally arranged for Shi Peipu and the boy to travel to France. There the French arrested both Boursicot and Shi Peipu as spies, and the trial revealed to Boursicot, after twenty years, that Shi Peipu was a man. These events became the basis for *M Butterfly*, which won a Tony for Best Play in 1988.

Such a tale raises all sorts of questions, and one is: How could Boursicot have failed to see Shi Peipu's maleness in his face? We normally recognize facial gender at once. We use secondary tipoffs, like hair length and makeup, but we hardly need them. In one experiment, researchers showed volunteers 185 photos of women and clean-shaven men, all in shower caps to hide their hair. Subjects were 96 percent correct in guessing whether the face was a man's or a woman's. Other studies have yielded similar results.

The skill is central, since facial lines define the sex that attracts us. We need to know them to perpetuate the species. Indeed, at puberty male and female faces grow more disparate, and in old age they swing back toward similarity again. The genes exaggerate the distinctions during the childbearing years.

Identifying gender by face is so important that evolution seems to have made it automatic. Hence few can articulate the differences between men's and women's faces. The feat has long stumped scientists, and only recently have they gained some insight into it.

The clues lie scattered across the countenance.

Men in general have craggier features. Their brows and chins jut out more. Their foreheads often slope more steeply, and their eyes lie in deeper sockets. They have longer cheeks and greater overall depth of face. More hair follicles dot their faces, which can make their skin look coarser, especially in older men.

Women, on the other hand, have smaller faces, usually about four-fifths the size of men's. In addition, their faces look more childlike. They seem wider and their eyes appear much bigger. Audrey Hepburn has relatively larger eyes than William Holden. The tissue around them is more sensitive to changes in blood circulation and darkens more quickly, an alluring effect women heighten with mascara. Their eyelashes are longer and thicker. Their eyebrows are thinner and grow sparser with age, while men's get weedier.

The nose also sets the sexes apart. Women have smaller, wider, and more concave noses, like children's. Men's are bigger and more protrusive, possibly because men need a larger respiratory system overall, from lungs to breathing passage to nose. In one study which showed noses in isolation, people identified male noses better from the front and side, female noses from the three-quarter view. Researchers speculated that all noses look somewhat male from the front, and that the three-quarter view better revealed the diagnostic bridge.

Women have other signal features. Their mouths are smaller and their upper lips relatively shorter. Their cheeks stick out more than men's, because of both their smaller noses and their extra layer of fatty tissue there.

Women's faces are smoother than men's. Not only are their facial muscles smaller, but their facial fat hides them better. Hence, their minor facial movements are less detectable. Men seem to have more mobile faces, and some researchers think we associate facial mobility with masculinity, facial stasis with femininity.

But this effect obtains only for slight movements, and in fact women are more facially expressive overall. Their faces respond more to highly charged material. They report feeling stronger emotions, can better match their faces to expressions in photos, and display more happiness and excitement in interviews.

There is no litmus test for telling men's faces from women's. Features like nose length and cheek protrusion overlap between the sexes and are not dispositive. Vicki Bruce of the University of Stirling in Scotland masked parts of the face and tested people's ability to tell male from female. "You get graceful degradation," she says. "If you cover the eyebrows, the ability doesn't disappear. If you remove information about the nose and chin, the ability doesn't disappear. It looks like the human system makes use of all the pieces."

Intriguingly, one study showed that people can not only handily separate faces into male and female, but also easily rate faces on masculinity or femininity, a rather different task. That is, they can say, "This is a masculine-looking face, but I know it's a woman's." We sense aspects of the male pattern even though we round off to female.

One group of investigators sought to fuse numerous facial variables into a single formula for distinguishing male and female faces. Though somewhat successful, these researchers concluded that the great lesson of the endeavor had been the difficulty of deriving such a measure.

Men have posed as women repeatedly onstage, and not just in the time of Shakespeare and Molière. The male *onnagata* of Kabuki have played females since 1629, when the shogunate banned actresses because their prostitution had led to jealous feuds among the nobles. Some *onnagata* lived as women offstage as well, and one once wandered absentmindedly into the female section of a bathhouse. Police arrested him.

The female impersonator Julian Eltinge (William Dalton, 1883–1941), who had soft features, large eyes, and chubby cheeks, became rich from his talent. Ruth Gordon said that offstage he was "as virile as anyone virile." He liked beer and horse-racing, cultivated a reputation for bar-brawling, and repeatedly stated that he was in it for the money. Many performers, like Charles Pierce and Jim Bailey, mimic female celebrities onstage. And the whole plot of *The Crying Game* hinged on Jaye Davidson's duping of the audience.

Women have acted male parts on stage too, though less often. Sarah Bernhardt (1845–1923) made a sylphlike Hamlet, and Sarah Siddons, Judith Anderson, Eva Le Gallienne, and over fifty other actresses have also played the disturbed Dane. Peter Pan has almost become a female bailiwick, dominated by actresses like Mary Martin and Sandy Duncan. In Japan, the actresses of the all-female Takarazuka Revue pose as men and woo their women ardently, as Japanese men notably do not. About 2.5 million fevered fans pay to see them each year, and 95 percent are female.

But the most fascinating impersonators are those who successfully live as the opposite sex. Profane, cutlass-wielding pirates Anne Bonny (c. 1695–?) and Mary Read (d. 1721) dressed and passed as men; they apparently kept this secret even aboard ship, though it is unclear how. Actress Eliza Edwards (d. 1833) was deemed a great beauty and the women who cared for her in her death throes were amazed to find she was a man. The Chevalier d'Eon (1728–1810) was a male general who retired at forty-nine, then convinced everyone he had been a woman passing as a man, and posed as a female for the last thirty-two years of his life.

The secrets of long-term deceptions like these probably twine art with transsexual will and sensibility. But stage performers cross the gulf with a variety of tricks. They don the right clothing, alter hairstyle, bind and pad their bodies. They use makeup wisely. Eltinge accented his lids with blue, thickened his lashes, and made his eyes almond-shaped. In general, cosmetics that make the nose smaller, the skin smoother, the eyes bigger, and the lips redder will make a man look more like a woman, and vice versa. Good impersonators also mimic voice, gait, and gesture. The actresses in the Takarazuka Revue, for instance, learn men's mannerisms from the films of Cary Grant, Tyrone Power, Sean Connery, and Kevin Costner.

Like ventriloquism, magic, and conventional acting, this feat demands years of practice. We underestimate the power of long-term study, and a person who flourishes enough tricks at once can sometimes conjure a true vision.

Three
3

Emblems of Self

WHEN anthropologist Edmund Carpenter handed out the first Polaroid portraits to tribesmen in the New Guinea highlands, the gifts bewildered them. They weren't used to two-dimensional images and Carpenter had to explain, first pointing to a feature like the nose in a photo, then the nose of the person. As their faces leapt out of the grays, these people ducked their heads and turned away. Some repeated this reaction several times, but then they often stared at the picture, mesmerized, their stomach muscles clenching with tension. Many slipped away, photo pressed to chest, seeking solitude in which to study it. They were confronting their visual images for the first time and the shock was deep. But later some wore the Polaroids on their brows, as proclamations of self.

The face is a deep surface, both skin and mind, and there are other peoples who resent its capture by outsiders. American Indians called early photographers "shadow-catchers," men who seized the symbol of death. In 1901 an old aborigine of central Australia complained that Sir Walter Baldwin Spencer was shooting photos "to extract the heart and liver of the black-fellows." The phrasing is extreme, but the feeling resonates, for who doesn't feel a suck of violation if an utter stranger snaps one's photo?

Images of the face differ radically from those of hands, say, or landscape. They are full of embedded meaning, and a certain uncanniness can wriggle into them. They can be taboo. The portrait face struggled toward likeness for centuries, and the face of God has long tantalized the faithful. They can be icons, mirage superidentities. They can be caricatures, subidentities which transcend jest and hold surprising secrets for face recognition. They can be fantasy substitutes. Masks make us supernatural, spur amours, and hide and reveal our selves. And where they are blank, where the face is unknown, the mind drafts the imagination to supply it.

Even the face in the mirror is not quite what it seems.

A History of Mirrors

We are experts on ourselves. No one else knows our thoughts, yens, or history like we do. Yet others have the edge on us in one regard: our faces. They see the play of our expressions and we don't. Rochester tells Jane Eyre of the "bloom and light and bliss" in her face, which neither she nor the reader had sensed. To gain such knowledge, we use the mirror. It is an everyday out-of-body experience that shows us our face.

Yet mirrors have always been a battleground of insight and illusion. The image they present is both true and deceptive. It is a sharp, vivid picture, but the man or woman in the mirror does not exist. Even so, at one time or another almost everyone has reacted to this photon clone as if it were a distinct person. We wink, smile, nod, salute, and watch the mirror-twin respond in kind.

Before the Renaissance, most people never saw their faces clearly. The earliest known mirrors are obsidian disks from Çatal Hüyük in Turkey, dating back to 6500–5700 B.C. The Egyptians devised the first metal ones, of copper, in the First Dynasty (2920–2770 B.C.), and the resourceful Olmecs used magnetite, hematite, and iron pyrite. But bronze was the surface of choice. The Greeks and Etruscans favored it, and the Chinese devised artful bronze mirrors with cosmic symbols on their backs. To ward off evil spirits, the Chinese often buried people with a mirror facing upward and in 296 B.C. King Ai of Wei went to the grave with several hundred of them. Mirrors had other special powers. Ground up and mixed in potions, they eased the pains of childbirth and cured heart problems.

From the start, mirrors have held witchery. The clairvoyants called scryers appeared early and everywhere. They discerned faraway or future events in mirrors, or sometimes in lakes, crystal balls, cups of liquid, the oiled nails of a youth. Scryers exploited the tantalizing blur of mirrors, suggesting they could see images invisible to other mortals. The Aztec god Tezcatlipoca, or Smoking Mirror, was an all-powerful scryer who divined the fate of every individual in an obsidian plate. Merlin and Prester John also owned such magic mirrors. *Specularii*, gifted men with lesser instruments, roamed medieval Europe and often employed children, who could go blind from staring at glare. The Pawnee collected the blood of a badger and youngsters read their destiny in it by moonlight. And mystic Jacob Böhme (1575–1624) believed the scryer's gaze emitted spiritual energy which coated the mirror so it reflected celestial landscapes and could show past, present, and future.

The clarity of modern mirrors killed scrying. The Romans had actually invented the glass mirror, but its lead backing gave a muddy image, so the mirrors of chivalry were bronze. The

Venetians created the modern article. Around 1460 they invented the first transparent glass, and in 1507 Andrea and Domenico d'Anzolo del Gallo applied a tin-mercury amalgam to one side, creating the first truly detailed mirror. The invention spurred a lucrative industry in Venice, which veiled the method in secrecy and preserved its monopoly for over 150 years.

Colbert broke it in 1664, when he bribed Venetian mirror workers to come to France. Venice threatened their families and a fantastic sequence of plots and counterplots ensued, in which two artisans died of poisoning. But by 1671 France scarcely needed Venetian mirrors anymore and the technology was loose. To boost the new French industry, Louis XIV commissioned a famous corridor for Versailles: the Hall of Mirrors.

Mirrors can spy for us. Anne Frank noticed boys stealing glimpses of her in the mirror in school, and health club members peek at others in the wall mirrors. Semimirrors like windows are more discreet, since they yield a ghostly image one can pretend to gaze through. In Kawabata's *Snow Country* (1948), Shimamura secretly studies Yoko's reflection in a train window, watching her face pass over the mountains until it seems to merge with them. TV and computer screens are semimirrors. So are eyes, and lovers see their image in each other's pupils, enlarged by the convexity.

We search our selves in mirrors. The mirror was the symbol of knowledge at the Yamato court, and the Japanese often incised pictures of the Buddha onto its surface, so mortal and paragon could merge. Mirrors show the riddle we present to others and ask us to scry our souls, an exercise that is doomed and irresistible. We usually perform it alone, and faces only become fully alive in the presence of others. Yet mirrors let us treat our image as a separate individual, whom we cannot disturb by staring at. Watching, we become first and third person at once, viewer and viewed. Even when the face is utterly impassive, we can know it's thinking, *I know what it's thinking*, and enjoy a moment of mindreading.

Some people seek their very humanity in mirrors. In Yevgeny Zamyatin's dystopic *We* (1921), the cipherized engineer D-503 stares into a mirror trying to find his soul, and detects a glimmer. Roquentin in Sartre's *Nausea* (1938) finds less. "Obviously there are a nose, two eyes, and a mouth," he says, "but none of it makes sense, there is not even a human expression." Indeed, he wonders whether anyone can understand his own face. In one literary convention, reflection requires a soul. Dracula has forfeited his and can't see himself in the mirror.

Placed before mirrors, parakeets and Siamese fighting fish will battle, court, and even feed their images to exhaustion, while cats and dogs soon lose interest. When primates first see themselves in a mirror, it tricks almost all of them. They threaten or make friendly overtures, as if confronting another primate, and peek behind the mirror. Most monkeys never learn. They treat the image as friend or foe until they tire of it.

But apes come to realize it is themselves. In one famous experiment, Gordon Gallup of the State University of New York at Albany anesthetized mirror-trained chimps and painted odorless red lines above their eyebrows. When they awoke, he set a mirror before them. They all tried to touch their foreheads, and many then smelled their fingertips. The experiment proved the chimps recognized their images, since otherwise they would not have checked their own bodies. Gallup later found that orangutans, gibbons, bonobos, a few gorillas, and children over eighteen months show the same reaction.

Some chimps and orangutans go further. They use mirrors to examine their teeth, rumps, and other body parts that usually lie out of sight. They make comic faces, place vegetables atop their heads, even adorn themselves, draping vines around their necks. Why would an ape act this way? Is it playing with its image? Testing facial expressions? Enjoying control of the mirror-slave? Gallup concluded that chimps and orangutans were "self-aware," a term that made many scientists bridle. But such behavior is mystifying without some degree of self-awareness. The key questions are: How much, and of what kind?

We improve ourselves in mirrors. Dr. Henry Jekyll used one to witness his slide into Edward Hyde, but most people try to ascend. We shave, floss, brush, comb, apply makeup. Indeed, the sophisticated use of cosmetics requires mirrors. Compacts let women spruce up anywhere, and the portable mirror is ancient. Tsimshian Indian women of British Columbia hung a piece of slate around their necks, and when they felt an urge to see their faces, they licked it and held it at an angle to the sun. Venetian glaziers found they could apply their tin-mercury amalgam to any size glass, and pocket mirrors proliferated. Playwright Philip Massinger refers to them in *The City Madam* (1624). In Tokugawa Japan (1603–1867), people wore sleeve mirrors, which let them discreetly check their faces as they strolled down the street. French ladies carried mirrors near their hips. "Alas! what an age we live in to see such depravity as we see," sighed moralist Jean des Caurres in 1575, "that induces them even to bring into church these scandalous mirrors hanging about their waist!"

We enhance ourselves in other ways. In the mirror we toy with façade, smirking, smiling slyly or shyly, gazing deeply. Like actors, we experiment with expressions and try to assess them as impartial onlookers. The face becomes a soft medium, sculpted from within. Thomas Nashe (1567–1601) complained about women who practiced fetching glances in their mirrors, but it can go farther. People can seek control of total self-presentation and wind up as artificers.

Curiously, mirror self-improvement can occur almost reflexively. Some individuals constantly judge themselves in mirrors and quickly paste on alluring expressions, make mirrors their sycophants. Joyce's Gerty MacDowell might weep in private, but she "knew how to cry nicely before a mirror. You are lovely, Gerty, it said."

Even alone with a mirror, people trick themselves. The eminent evolutionist Robert Trivers notes that lying is central to animal communication, since it confers major boons. Therefore, he says, "there must be strong selection to spot deception and this ought, in turn, to select for a degree of self-deception." It happens because self-delusion makes lies more believable. If we've already fooled ourselves, we'll seem candid and even earnest, and thus fool others better. Litigators know they can argue a case more effectively if they persuade themselves the client is right, and salespeople routinely pump themselves up about the product. Research shows that many flight attendants who must politely handle an obnoxious passenger will first convince themselves they like him.

And self-deception can occur automatically. For instance, we unconsciously concoct and believe excuses for failures, a trait psychologists deem healthy. We tend to interpret ambiguous feedback positively, though studies show it is usually negative. Rage blinds us, probably to make us more powerful fighters. Know thyself, said the Delphic oracle, and the advice has limitations. We evolved for competition, not insight, and it's one reason we can seem enigmatic to ourselves. It's also one reason the mirror can fascinate us.

Indeed, it can be a pool of intoxicating self. Han Fei Tzu (280?–233 B.C.), the Chinese Machiavelli, said a mirror in which we don't see our flaws is like self-analysis which doesn't acknowledge our sins. But such mirrors abound. The mirror is the symbol and delight of vanity, and was once called a "flattering glass."

In Vonnegut's *Bluebeard* (1987), egotistical artist Dan Gregory has 52 mirrors in his studio, set in a wild array of angles. The young Balzac placed parallel mirrors in his Paris garret to savor endless copies of his face. In *Richard II*, the king swims in flattery, and after his downfall he likens a mirror to his toadies and smashes it on the ground.

Mirrors can speak. In Isaac Bashevis Singer's short story "The Mirror," the beautiful Zirel wiles away the hours studying her naked flesh. A demon inhabits the mirror, whom she can summon with her eyes, and he cajoles her into flying off with him to Rahab the Harlot's palace, where she meets a grim sexual destiny. A mirror, says the demon, is "a kind of net that is as old as Methusaleh, as soft as a cobweb and as full of holes, yet it has retained its strength to this day." In the original "Snow White," an almost reportorial mirror daily informs the queen of her supreme beauty, and she takes it so seriously that when its opinion shifts, she attempts homicide.

Mirrors have other odd powers. They flip the visual field, so a left-cheek mole moves to the right cheek. Hence, when Alice enters the looking glass, she finds a world of reversals and oddities. In Haitian voodoo, the lwas or loas, whose interaction runs the universe, are mirror-images of the personalities of the living. Indeed, reverse images of the profane world pervade voodoo rituals, cosmology, and sense of time.

Mirrors offer alternate worlds, in which a mimic of ourselves inhabits a realm like ours. When an Easter Islander first saw his face in a mirror in 1722, he quickly looked behind it, checking for this second universe. Alice fantasizes aloud about a Looking-Glass House before entering the mirror, and the young Jane Eyre sees a colder, darker sphere in the looking glass, where her image seems a lonely spirit, its eyes bright with fear. In *The Phantom of the Opera*, a mirror acts as the secret entry to Erik's very real world.

But the alternate world received classic expression in Ovid (43 B.C.–17 A.D.?). The water nymph Echo loved Narcissus, who spurned her. She hid in caves and leafy forests, but could not shake her love and wasted away to mere voice. Narcissus had rejected others, who prayed for vengeance. Nemesis heard them. One day Narcissus came to a pool, gazed at its mirror surface, and fell in love with his reflection. "The loved becomes the lover," writes Ovid, "the seeker sought, the kindler burns." Narcissus couldn't leave the pool. Every time he leaned down to kiss his image, it reciprocated. It loved him too, he thought, yet eluded him. When he finally realized the image was himself, he yearned to shed his body, to leave it behind on the bank so he could merge with the image. Finally, like Echo, he wasted away, leaving behind just a white-petaled flower with a yellow center.

The face in the mirror is everything, and it is nothing at all.

The Ultimate Image

"Nothing in the whole circle of human vanities takes stronger hold of the imagination than this affair of having a portrait painted. Yet why should it be so?" asked Nathaniel Hawthorne. He thought the answer lay in permanence, for we hardly fuss so over our face in a mirror.

Indeed, a mirror-image is life itself and vanishes constantly into the past, while a portrait can last thousands of years. Friends, descendants, utter strangers can view it and size it up. It becomes a wandering piece of reputation.

A likeness peels off a bit of the self. It is both shape and sense, and full of tug. Hence the face has been the most durable and significant image in art.

The earliest known representational art is the Vogelherd horse, a beautiful miniature ivory carved 32,000 years ago. But human likeness appears early. At the 26,000-year-old site of

Dolní Vestonice, near Brno in the Czech Republic, archeologists discovered a pair of human faces, one ivory, the other clay. Both droop on the left. A few yards away they also found the body of a woman of about forty whose bone disease would have made her face sag on the left. If portraits, they are the oldest on earth.

People make most history, and portraits have long had a historiographical aspect. For instance, they could show genealogy. Maoris carved faces and tiny bodies of ancestors on pou-pou ("po-po") posts, and placed descendants' faces between their legs. These posts remain valuable keys to their history, since the Maoris lacked a written language.

Totem poles could depict important deeds in a family's past, though they quickly weather and rarely last beyond a hundred years. One Haida pole in Alaska shows the nineteenth-century Russian priests who had tried to convert Chief Skowl and his tribe. The Haida used the totem pole to historicize the event and ridicule the defeated missionaries. Often, when one chief vanquished another, he published his victory on a pole, so all could mock the conquered foe.

The seventeenth-century Dutch commissioned portraits to freeze key moments in life, such as weddings and baptisms, and we compile similar histories today: album biographies. We photograph faces at graduations, parties, trips to Cozumel, and other notable events. We fix our image at certain times in life, as in yearbook photos. Parents record their children's growth, and journalists memorialize the famous.

We once fixed faces at death. Thus on the night of June 13, 1793, agents of revolutionary France came calling on Madame Marie Tussaud. They told her to go to the apartment of Jean-Paul Marat, whom Charlotte Corday had just stabbed in the bathtub where he worked, and cast his death mask. Tussaud later claimed coercion, but in fact throughout the Terror she regularly received blank-eyed heads from the guillotine and took wax impressions of them. She made death masks of Corday, Madame du Barry, and many Jacobins she had earlier modeled alive, including Robespierre himself. In 1802 she escaped to England with her waxworks, which later adorned wax museums. They owed their success to the morbid accuracy of their likenesses, for in an age before photography, they caught the contour of famous faces. Yet they rarely captured the self inside, and the term "wax figure" has come to connote lifelessness.

Occasionally a likeness acts as second self. In 1341 Jeanne de Bretagne and her daughter wished to pray for miracles at the shrine of San Diego de Compostela, but felt the trip would be too tiring and dangerous, and sent silver statues of themselves instead. A more lurid case involves artist Oskar Kokoschka (1886–1980). In 1912 he began an affair with Alma Mahler, which she called "one fierce battle of love." Though her prior husband, composer Gustav Mahler, was dead, Kokoschka was wildly jealous of him. His rages irked her, and in 1915 she married architect Walter Gropius. Kokoschka fixated on Alma, and painted her face on many portraits instead of the sitter's. In 1919 he had a Munich doll-maker build a life-size replica of her, anatomically complete. It may have shared his bed, and he had his maid garb it in fine lingerie and dresses. Kokoschka made some thirty drawings of it and at least two paintings. Finally he threw a champagne party which lasted till dawn, when he decapitated the doll in the garden and poured red wine over its head. It had cured him, he said.

In art, likeness has developed a reputation as rote and dreary business, face sapped of soul, the mark of the third-rater. But in times past it has been an achievement, since at least it implies truth. For about a millennium in Europe, from 400 to 1400

A.D., likeness almost vanished. Artists sought to show spirit rather than facial contour, an undertaking that slighted individuality and led to stereotyped, factitious images. With the Renaissance likeness crept back, and in Holland Jan van Eyck (d. 1441) created the first modern portraits. Most of the Dutch portraitists specialized in hard detail, showing maplike versions of the face in clear, cold light.

But the popularity of likeness has risen and fallen in an erratic tide, and it has usually dropped in times of poverty, ignorance, and aspersion of the singular.

Likeness has other enemies. Sitter vanity has always bedeviled artists. Ayn Rand placed a preposterously beautiful sketch of herself on book jackets. Should artists show patrons as they are or as they wish to be? Flattery is the easy course, the one that wins clucks and fills the coffers, and many choose it. "Nature herself is not to be too closely copied," smiled Joshua Reynolds (1723–1792), who did 650 portraits and was the best-paid artist of his time. The ever-pragmatic Andy Warhol often made several portraits of his sitters, hoping at least one would please them. In *Nausea*, Roquentin wanders the Bouville art museum gazing at portraits of local citizens. "They had been painted very minutely," he thinks, "yet, under the brush, their countenances had been stripped of the mysterious weakness of men's faces."

It takes daring to depict a simpleton as he is, and Francisco de Goya (1746–1828) had it. In *The Family of Charles IV* (1799), he presented the inbred royal family with dull, stunned faces, their dazzling garb highlighting their vacuity. Goya not only escaped their wrath, but the painting apparently pleased them. At least 300 nobles and 88 members of the royal family sat for Goya, willing to endure cruel accuracy for the chance at immortality.

Other sitters have been less tolerant. Stalin stood five-feet-four, but wanted painters to depict him as a tall man with hands that could uproot trees. Several failed, and he had them shot and their work burned. The artist Nalbandian showed him as an imposing figure with his hands clenched powerfully across his stomach, and finally satisfied the dictator. (Stalin once revised the official *Short Biography of Stalin* to add: "Stalin has never allowed his work to be marred by the slightest hint of vanity, conceit, or self-adulation.")

Stalin was an extreme client, but the commissioning of portraits taints them. A sense of compromise festoons the genre, as it does the authorized biography. Not counting herself, the artist has two audiences: the sitter and the public. Whom does she serve? Does she give the world truth or confection? Even these questions are misleading, for here as everywhere, the most effective flattery is subtle. It enhances but remains credible. The sitter's best friend may seem the public's friend as well.

Role can warp likeness. Old Kingdom pharaohs gaze out as idealized god-kings, almost clones. Likeness appeared suddenly in the Middle Kingdom, whose rulers seem individual and alive, but the divine façade descended again with the New. Salvador Dali's *Mae West's Face Which Can be Used as a Surrealist Apartment* (1933–1935) depicts West as a pure creature of film. Her face is a movie screen and her hair tumbles down like its curtains. She has a surface, composed of eyes, nose, and painted lips. Yet like the basic movie illusion, her face also recedes into a third dimension, revealing a near-vacant room where her lips form a sofa. (Dali later won commissions for full-size lip-sofas.) The portrait suggests her likeness, but it is really a study in public image.

Portrait artists can also sacrifice likeness to general preconception. When Benjamin West painted *Benjamin Franklin Drawing Electricity from the Sky* (1801), he depicted the canny elder we see on the hundred-dollar bill. But Franklin was around forty when he defied ultrahigh voltage with his kite. His image only became fixed in the public mind when he was older, and West chose familiarity over realism.

At least since Petrarch, artists have painted portraits for moral instruction. In the early Renaissance, they sought to enhance the virtue in their subjects, to show the paragon lurking in the ordinary citizen. In 1543, after Sebastiano del Piombo completed a portrait of Venetian humanist Claudio Tolomei, Tolomei wrote the artist praising its inaccuracy. The portrait, he explained, would goad him to attain the perfection del Piombo had graced him with. He added that he saw both himself and del Piombo in the work, and the sight of the latter's genius would spur him to honor and glory.

Portraiture has always hinted at eternity. It promises a laugh on oblivion, conquest of the deathly void. Indeed, the ancient Egyptians aimed for literal immortality. Statues were magical and, once decked with name and title, housed the sitter's spirit forever.

But the typical portrait preserves information. The soft, transient face finds a second life in perduring material. In *The Epic of Gilgamesh* (c. 2100 B.C.), Gilgamesh has artisans make a posthumous statue of his friend Enkidu. Early Romans made many death masks, and later ones carved portraits on their tombstones. And artists like Goya have attracted sitters who want future centuries to gaze on their faces.

It is a curious gift to posterity. It can be valuable. We know what Nefertiti and Julius Caesar looked like because busts of them survive. Martin Droeshout's picture of Shakespeare on the First Folio (1623), though probably drawn from memory, is our only image of him and of great interest.

But posterity usually responds with a yawn. Who cares whether *The Blue Boy* accurately depicts Jonathan Buttal, the son of an ironmonger? As Walter Benjamin observed, "The portrait becomes after a few generations no more than a testimony to the art of the person who made it." Goya's noble sitters may be making a perky cry from the grave — "I live on!" — but we only care if their faces flicker with something of ourselves. Identity survives in universality.

Philosopher Jeremy Bentham (1748–1832) made a more intriguing attempt to persist. After his death, he told his heirs, they should save his skeleton, reconstruct a body decked in his own clothes, and mount a wax head of him atop it. Today this effigy sits slouched in a glass cage at University College, London, Bentham's jowly face looking upward as if still curious about the world. Bentham hoped other thinkers would follow his lead, but so far he sits alone.

Some portraits honor the transfer of identity over time. Zen Buddhist *chinzo* show the master along with a personal dedication, such as "From Mind to Mind." The student received one on attaining enlightenment, and it represented wisdom crossing generations. The Roman double herm was a pair of busts fused at the back: a Greek like Aristophanes joined to a Roman like Terence, his heir and implicit equal. The apotheosis sculptures of the Huastecs in Mexico present the monarch on one side and a baby face or death's head on the other, probably his son or father. They show links in the chain of royalty.

Louis Daguerre discovered how to fix a photographic image in 1837, and portraits in the new medium appeared soon after. Posing was agony at first. In 1839 Lord Brougham sat for half an hour in the hot sun and called it the worst experience of his life, but the time soon dropped to minutes. We have a daguerreotype of Balzac in 1842, liquid eyes gazing left, palm pressed to his chest, the faithful philanderer. The early portraitist Julia Margaret Cameron (1815–1879) developed a soft-focus style that etherealized subjects, making their faces gaze out as if from a dream. But she was an illusionist, and the typical sitter, like Daniel Webster (1852), shows a skin-pore detail utterly different from that in paintings. It is hard not to feel that, with the photo, we pass a continental divide in historiography. We see the real texture of the past.

Armed with this verisimilitude, photographers fanned out to document the faces of the world. Roger Fenton (1819–1863) brought back faces from the Crimean War, and the energetic Mathew Brady (1823–1896) and his group memorialized the Civil War. W. H. Jackson (1843–1942) traveled west in the 1870s and photographed Indians, a delicate enterprise given their belief in the soul-stealing nature of the camera.* Brassaï (Gyula Halász, 1899–1984) roamed the Paris underworld of the early 1930s, recording cafe couples, singers, dancers, opium smokers, strippers, and prostitutes, and Diane Arbus (1923–1971) presented a gallery of dwarves, nudists, homeless, and cross-dressers.

In 1888 George Eastman invented the first "box camera," which he called a Kodak. (He liked the name because it was short, memorable, and hard to mangle.) The buyer paid $25 for the device, which came with 100 frames of film, and held it tightly to his chest to snap photos. When he finished the roll, he mailed the whole camera back to Rochester, where Eastman's company developed the film for another $10. The shrewd inventor had divorced picture-taking from film-developing, and

---

*Camera-toting anthropologists eventually learned to compensate their subjects. "The Caduveo had perfected the system," says Claude Lévi-Strauss. "Not only did they insist on being paid before allowing themselves to be photographed; they forced me to photograph them so that I should have to pay."

this strategy revolutionized the industry. Eastman knew the profits lay in the film, and in 1889 he began selling it separately. His work launched a craze in the 1890s. Anyone could now be a portraitist. Even the Dalai Lama bought a box camera. Warned the *Hartford Courant*, "Beware the Kodak. The sedate citizen can't indulge in any hilariousness without incurring the risk of being caught in the act and having his photograph passed around among his Sunday School children."

The camera was faster, cheaper, and truer than the brush. It took over likeness and forced art into boutique specialties. Scores of schools emerged. "A head is a matter of eyes, nose, and mouth," declared Picasso, "which can be distributed in any way you like; the head remains a head." From around 1906 on, his faces grew more and more jumbled and quixotic until they finally became simple markers of his mood. Francis Picabia's dada *Portrait of Marie Laurencin* (1917) tweaked the whole genre, showing a motley of mechanical parts with no link to Laurencin herself. As Tristan Tzara liberated language from meaning, he freed portraiture from the face.

Painters had always infused portraits with aspects of themselves, but now tendency became need. Modern portraitists like Francis Bacon (1910–1992) worked up distinctive styles which often entombed likeness. Alberto Giacometti (1901–1966) sculpted figures as anonymous knurled poles, yet claimed to seek likeness. ("Only when long and slender were they like.") As Modigliani sank into alcoholism and drug addiction near the end of his thirty-five-year life, every sitter's face became the same: long sloping nose, almond eyes, often tilted head. These paintings became the trademark Modigliani, the ones forgers mimic. In the seventies Andy Warhol mannequinized his subjects, whitening their faces with a powder that vaporized wrinkles and indeed all skin detail. The countenance became a blank, except for eyes, eyebrows, lips, and the shadow of nose and chin. They look so similar that even today many portraits in his workshop remain unidentified.

Meanwhile, studio photographers made capturing likeness a small industry and cameras spread everywhere. Kodak unveiled the Instamatic in 1963 and sold almost 60 million over the next decade. Today's automated cameras do almost everything but frame the shot. Each year, Americans alone take more photos than there are people on earth. One result, as many have noted, is a wilderness of potential found art.

Photography seems a Diogenes technology, eminently truthful, and in places like India people still arrange marriages based on it. But the camera captures light, not actual surface, and professionals command a raft of subtle tricks. For instance, low-grain film or longer exposure smoothes the skin. Lighting a face from above at 45 degrees right or left yields the Rembrandt pattern, a classic look, while lighting the same face from above head-on gives the butterfly pattern, which glamorizes the sitter. And of course retouching in the darkroom removes wrinkles and other blemishes.

Computers now let us manipulate photos totally, pixel by pixel. Already they have enhanced photos of Elizabeth Taylor, altering her face shape and skin tones, and the technique is becoming routine. Modern Ayn Rands need not resort to drawings to fool the reader, and indeed the portrait of myself on the jacket of this book may not quite be me.* The link to reality will soon vanish from the photo, as it has from the movie screen, and we will return to the world of the salon portrait, with all its conflicts of honesty and audience loyalty.

---

*It is.

Mao and Marilyn

In the *Popol Vuh*, godlike Seven Macaw brags, "I am like the sun and moon for those who are born in the light, begotten in the light. It must be so, because my face reaches into the distance." However Seven Macaw deludes himself. His face does not extend as far as he thinks, and others soon overthrow him.

But the faces of more prehensile rulers stretch to every corner of the land, and peer out of every cranny. The face is the classic power icon. It appears on coins, currency, medals, stamps, billboards, public busts. It can reach grandness, as at Mt. Rushmore and Easter Island, and grandiosity, as in the colossal images of Stalin, Mao, Kim Il Jung.

All icons fuse a face with an idea. They do not explore identity, but distill and create it. However these countenances are special. They are icons of authority. Mao stares out over Tiananmen Square and he is the image of an overlord. His face evokes a visceral response from some knot in one's being.

The Japanese phrase *kao ga kiku* means "to be influential." But it translates literally as "to exert face," and power icons exert face in numerous ways.

In primitive cultures they can throb with personal force. The Kipchak idols (11th–12th cen. A.D.) near the Black Sea, stone figures of nobles that rise above burial mounds, gaze sternly at the world they once inhabited. The older *moai* on Easter Island brim with scorn, and the famed Olmec heads, like thick-lipped, 18-ton Colossal Head 5, stare out with serene ferocity. Such faces send a clear message. Before the calm power of Colossal Head 5, a frisson of dread is appropriate.

Other icons merge the ruler with legends or gods. Egyptians carved the Great Sphinx at Gizeh from a natural outcrop of rock around 2550 B.C., and though Islamic authorities destroyed its nose 500 years ago, it retains an opaque splendor. It likely depicts the pharaoh Khephren, in idealized form. The Greeks viewed the sphinx as a riddler, an enigma, but the Egyptians were simply fusing Khephren with prehistoric notions of the king as potent man-beast. Ceremonial objects show pharaohs as lions trouncing their foes.

The Khmers of Cambodia featured a god-king cult, and Jayavarman VII (ruled 1181–1219) believed himself an incarnation of the Buddha. A ceaseless builder, he created the great Bayon temple at Angkor, a "world-mountain" representing the center of the universe. He adorned the Bayon with face-towers, where behemoth stone faces still stare out from the crumbling façade. They merge his own countenance with the Buddha's and, lids closed in contemplation, they preside over the temple complex and haunt it.

An icon can stress office. A stela at Tikal (c. 380 A.D.) depicts the tiny, drowsy face of Curl Nose sinking into his regalia as if it were quicksand. The emblematic portrait of Queen Elizabeth I, attributed to Crispin van de Passe (c. 1596), shows her aswim in majesty. Her head pokes out of a great wheel of ruff, above billowing robes, and she stands between pillars as ships bob in the distance. Her face is blank, anonymous, inconspicuous—almost the last thing we look at. She is more queen than Elizabeth.

Coins and currency weld the ruler's face to wealth, the goal of much earthly striving. They also make him ubiquitous. Lydia in Asia Minor invented coins between 620 and 600 B.C. and nearby Greek colonies quickly copied the idea. Persia placed the first portraits of kings on its gold diracs and silver sigloi. Since raised images wear away, maiming full faces, mints learned to show profiles. The Greek states stamped coins with gods like Dionysus and Pan, and avoided living persons. By the Hellenistic age, however, rulers were declaring themselves gods and mooting the distinction. Ptolemy I of Egypt (ruled 323–285 B.C.) placed his own face on a tetradrachm around 300 B.C. and other kings eagerly followed.

In Europe, the monarch's face was often the only one known nationwide, because it appeared on coins. In 1791, when Louis XVI fled north disguised as a valet, musket-wielding peasants captured him in the small town of Varennes. A postmaster had recognized him from his currency. Anyone else would have slipped through.

The lure of the coin-icon can intoxicate. In 1925 Martin Coles Harman bought Lundy Island, off the British coast, believing it independent. He declared himself its sovereign and issued a bronze coin called the half-puffin, with his face on the obverse. He had joined the ranks of Augustus, Henry VIII, and Washington, and when English courts rejected his claim of autonomy, the coins became curiosities.

The first adhesive stamp, the "Penny Black," appeared in Britain in 1840 and it too bore a ruler: Queen Victoria. Other living monarchs soon followed, such as Leopold I of Belgium in 1849 and Louis-Napoléon of France in 1852. These stamps showed heads in a regal profile taken directly from coins, often as busts. When the first American stamps debuted in 1847, they depicted George Washington and Benjamin Franklin; U.S. stamps have never displayed a living president. In Chile, for some reason, almost every stamp issued until 1910 showed Christopher Columbus.

A collected stamp is a gift to the government, and thrifty states try to swell collections. So stamps have diversified in Carboniferous riot, showing flowers, lighthouses, gems, landscapes, birds, fruit, almost anything. Faces remain a staple, but they have shifted from political to cultural icons. On its first stamp, the new Black Sea nation of Abkhazia (once Colchis, land of the Golden Fleece) featured Groucho Marx and John Lennon, and yuppie catalogs were promptly hawking it. Other cash-hungry nations soon followed: Tanzania (John Lennon), Chad (Jackie Onassis), and volcanic Montserrat (Marilyn Monroe). In stamps, variety is an economically stable strategy.

The best-known modern power icons are totalitarian and saturational. Jayavarman VII might fuse himself with a deity, but these rulers seem to become one.

In Zamyatin's *We* (1921), the face of the dictator called the Benefactor is always hard to discern. Even when D-503 meets him, his countenance seems lost in mist. The author never plausibly explains this fogginess of face, but it serves his purposes. The Benefactor is impenetrable, beyond human. The lesson is: Opacity is power.

Two decades, several dictatorships, and millions of lives later, Arthur Koestler handled the ruler's face differently in *Darkness at Noon* (1941). Pictures of No. 1 hang over Rubashov's "bed on the wall of his room—and on the walls of all the rooms next to, above or under his; on all the walls of the house, of the town, and of the enormous country." No. 1 is omnipresent, beyond human. The lesson is: Ubiquity is power.

George Orwell drew on both books. Icons usher us into the world of *Nineteen Eighty-Four* (1948). In the second paragraph, Winston glances down the hall of his apartment building and sees a vast Stalinesque face, over a yard wide. Descending the staircase, he meets this visage on every landing and its eyes seem to follow him. Beneath lies the famous legend: BIG BROTHER IS WATCHING YOU. Outside, he sees the face on house fronts and at every corner, and its eyes bore into him. "On coins, on stamps, on the covers of books, on banners, on posters, and on the wrapping of a cigarette packet—everywhere. Always the eyes watching you and the voice enveloping you. Asleep or awake, working or eating, indoors or out of doors, in the bath or in bed—no escape. Nothing was your own except the few cubic centimeters inside your head."

At twenty-one, Orwell had been colonial police chief in the Twante region of Burma, an area of 200,000 people, and had run an agency that spied on potential troublemakers. He had been Big Brother himself, and even after he left the East in 1927, the accusing countenances of the Burmese haunted him. "Unfortunately," he wrote, "I had not trained myself to be indifferent to the expression of the human face."

Big Brother lives today mainly in Islamic countries. The face of Turkmenistan president Saparmurad Niyazov appears in every office, shop, and school in the land, and the government tells citizens that if they criticize him, their tongues will fall out. Throughout the squares and souks of Damascus, the mega-face of Hafiz Assad stares down, smiling, a friend to all. Saddam Hussein actually resembles Big Brother. His face gazes out from museums, oil refineries, and town entrances, and every day it fills a rectangle three columns by ten inches on the front page of the Baghdad paper. A third of all TV time—some seven hours a day—focuses on him. By size, ubiquity, and position, his face takes on overwhelming meaning.

It's egomania, of course, but strategy as well. Such rulers seek to burn their probing eyes into every mind in the country. Multiply the face endlessly and insistently, and the very idea of it fires the gaze neurons. It has branded the consciousness.

A cultural icon is a mythic figure, a modern Aphrodite or Loki, an image on a grand and simplified scale. Elvis Presley, Marilyn Monroe, Michael Jordan, W. C. Fields, Robert Redford, John

Wayne, Martin Luther King Jr., Charles Manson, Albert Einstein, Mohandas Gandhi—such faces are instantly evocative. Some are simple epitome, like Adolf Hitler's, perhaps the most iconic of the twentieth century. They put a face on abstractions like evil. But most are also idols.

In ancient times, death and dimming memory might change a person like Agamemnon into myth. Today social distance and P.R. achieve the feat more quickly, and the great modern icon workshop has been Hollywood.

In the early days of film, studios shrank from icon-making. They feared the fame of actors would hike salaries, and hence kept them anonymous, so the public had to refer to favorites as the "Biograph Girl" and "Little Mary." The industry had created that rarity: the nameless icon. But in 1910 Universal Studios founder Carl Laemmle signed the Biograph Girl, revealed that she was Florence Lawrence, and publicized her shamelessly. The star system was born.

It quickly took over. Actors' fees did soar, but overall revenues outpaced them. Indeed icons of glamour—a term that can also mean "spell" or "enchantment"—made some studios rich. For instance, as early as 1914 Fox promoted Theda Bara (Theodosia Goodman, 1890–1955) as a Bedouin bedroom witch, an erotic vampire or "vamp." The term became a byword and she, along with cowboy icon Tom Mix, vaulted Fox to the top tier of studios.

A true star was a kimberlite, but also an image, a tissue of information, and studios boldly manipulated it. Ritualistically, almost symbolically, they rechristened actors with sprightly names. Makeup artists, hairdressers, and cinematographers beautified actors' faces, and they themselves often learned magnetic personas they could assume at will. Publicity departments chaperoned their identity offscreen, inventing bios, staging romances, and quashing news of crimes and other sordidness. And the studios plugged actors into the same role endlessly, to strengthen the iconicity. John Wayne seems like a hero partly because he played nothing else.

Such labors made people like Marilyn Monroe into commodity faces, mass-produced, globally familiar. Andy Warhol commented on it in his *Marilyn Diptych* (1962), which multiplied one image of her face fifty times, half of them in color and half in black and white, and tarnished the latter. Her life, he noted, was split between the chromatic and the sublunary. But every icon's was, to a degree, and Monroe herself came to hate people telling her she was beautiful.

Most icons convey a succinct idea, but some, like Jackie Onassis, can be fairly complicated. Wayne Koestenbaum, in his amusing *Jackie Under My Skin* (1995), asks such questions about this icon as: Was she a housewife? What do Jackie's sunglasses mean for us? Is icon Jackie the Medusa? What do Jackie and Elizabeth Taylor have in common? Is icon Jackie a freak or a madwoman? He is analyzing a silhouette, stretching it, tweaking it, trying to tease it into full-face.

As he plainly knows, the task is futile. Icons are almost masks. They fuse a potent notion like glamour with a face, and thus simplify and intensify both. The icon becomes a defining distillation, a richly human symbol, and it reaches deep into the mind.

The Face of God

In Archibald MacLeish's play *J.B.* (1959), Mr. Zuss and Nickles stand aloft commenting snidely on the action, but occasionally don masks to play God and Satan. Mr. Zuss wonders why they have to wear the masks. "You wouldn't play God in your face would you?" says Nickles.

What does the face of God look like? What *is* God? Does God have a face? If not, should we show him with one anyway? Should we display a coarsened and falsified image of God so people can comprehend him more directly?

These issues have the worst features of the will o' the wisp and the hornet's nest. The assumptions shift ceaselessly, real-world pressures make the reasoning laughable, and emotion runs high.

Some cultures with near-human deities have simply ignored the problem. Liberians made death masks of heroes and worshipped them as gods. They wore these masks into battle, where the feel of the hero-god's face hiked their adrenalin. Some scholars think the Olmec heads show actual chieftains, and on Easter Island, people bowed down to the stone faces of ancestor-gods, each distinct and probably based on a real person.

In the Tantric tradition, depiction is crucial. Worshippers must imprint a god's picture on their minds, grow so intensely aware of it that, as one scholar says, "the entire image glows within one, every detail alive with profound meaning, and the echoes of the deity within one's own being begin to stir." Icon spurs inner presence. Though some Tantrists make pictures they later discard as approximations, this is a luxurious embrace of idols. Most creeds are more ambivalent.

Polytheism staggers without images, since they help tell the gods apart. They serve other purposes too. Philo of Alexandria spoke of the urgent need of people to behold God, to see God's face. Everywhere, idols receive offerings, give people a countenance to pray to, and, we are told, heal the infirm. Indeed, they are so powerful we pygmalionize them and they wink, whisper, shed tears.

In the classical world, some felt the gods physically animated the statues of them. Their faces could smile, sag, glower, and otherwise communicate facially. They also acted as oracles,

murmuring advice into petitioners' ears. Dream interpreter
Artemidorus of Daldis (fl. 185 A.D.) said that if a statue of a god
smiled at one in a dream, good fortune was near. At times the
icons moved farther afield. A bronze statue of the hero Pelichus
liked to step down from its plinth and take a stroll.

But critics like Xenophanes and Lucian held that idols were
dead shapes. They pointed out that ordinary materials composed
them, which artisans might just as well have made into footbaths.
They laughed at the contrast between the god's power and the
statue's helplessness. Birds fouled them and they might rot, rust,
and erode under the rain. Arnobius (fl. 220 A.D.) told idolaters,
"Do you not see that newts, shrews, mice, and cockroaches,
which shun the light, build their nests and live under the hollow
parts of these statues?" And they cited human abuses. People
bathed the icons, and even chained them up. In Arcadia, says
Pausanias (2nd cen. A.D.), a veiled statue of Aphrodite stood with
fetters on her feet. Sophocles tells how, just before the fall of
Troy, its gods left bearing their statues on their shoulders. With
their images in chains, the gods were less likely to slip off.

Hindus make profligate display of their gods. At the temples in Madurai, they rise tier after tier, in the thousands. Unlike
the Greco-Roman gods, with whom some scholars discern
common roots, they often boast multiplied features like six arms
to set them apart from humans. At Harappa and Mohenjo-Daro
(c. 3000–1600 B.C.) archeologists have found three-faced gods.
Four-faced Brahma gazes at each compass point simultaneously.
At Elephanta stands a magnificent statue of Shiva with three
faces, by turns terrifying, feminine, and calm. Perhaps the most
famous god-image is Shiva in a whirling, multiarmed dance,
whose face yet retains serenity, the light of a faint smile. Sophisticated Hindus deem such portrayals pap for the masses, and
note that the gods exist on a higher, more abstract plane.

The controversy over icons looms larger in religions like Christianity, Buddhism, and Islam, whose central holy figures walked the earth in human frames.

The Second Commandment says, "Thou shalt not make unto thee any graven image, or any likeness of anything that is in heaven above, or that is in the earth beneath, or that is in the water under the earth." Read literally, this edict bans all representation and would leave us a world of De Stijl and arabesque. But the Israelites did not take it entirely seriously, and the Bible itself cites with approval pictures of beasts and cherubim.

Yet early Christians decried pictures of Christ as paganistic. Tertullian (160?–230? A.D.) denounced all divine images as falsity, hateful to God. Eusebius (260?–340? A.D.), bishop of Cesarea, told Constantine's sister that the crucified Christ's face shone like the sun, so glorious that even his disciples could not gaze upon it. No one could paint such a face, he insisted, and no one should try. The Council of 754 held that, since Christ embodied both man and God, a picture of him either included the illimitable Godhead, thus limiting it, or blasphemously excluded it.

But desire to see the divine face ran strong. John of Damascus (675–749), who felt we have an inborn yen to view God, stated the main pro-icon arguments: Images give the illiterate a sense of Christ. He had appeared in the flesh, and one could render him thus. The icon was merely a way of perceiving Christ, in Paul's words, "through a glass, darkly." John argued for wonder-working icons too. Peter's shadow had scattered demons and healed the sick, he pointed out, and an icon is more like a person than a shadow is. The Second Council of Nicaea (787) overruled the prior ukase, largely because icon-wrought miracles had grown so popular.

Despite later outbreaks of iconoclasm, especially during the Reformation, this decision had far-reaching effects. In the

Greek Orthodox Church, it led to icons which shimmered with divinity, received worship, had their own feast days, and routinely performed marvels. These images inspired ex-seminarian Joseph Stalin to drape the Soviet Union with pictures of himself.

In the West, the Church stressed the miraculous power of relics, like pieces of the True Cross, and such icons did not appear. But the Nicaea edict made the crucifixion one of the most familiar images in the Western world (and from the fourteenth to sixteenth century, many churches featured wondrous Christs with secretly movable arms and heads). And it led to some of the world's great art: Michelangelo's *Pietà* and da Vinci's *Last Supper* head a brilliant list.

Yet art means artistic license. Though da Vinci and Rembrandt took their models for Christ from local Jewish ghettos, others have been less scrupulous. In Nazi Germany Christ developed blond hair and in Africa he was black. Mexican artists favor Christ in agony, often with jets of blood spurting several feet from his wounds. Virgins have worn crowns, jewels, and precious robes. And this range is narrow compared to the spectacle in Buddhism.

Gautama Buddha (563?–483? B.C.) tried and failed to ban religious images, and his own now dots the Asian landscape. No other person has inspired such sculptural treasures. At Ajanta in India, 29 caves burrow into a semicircular cliff and each teems with sculptures and paintings of the Buddha, most executed in a single fifty-year burst in the fifth century. Japan boasts magnificent statues like the Todaiji Buddha at Nara (751) and the Seated Amida (13th cen.) near Kamakura. The temple of Borobudur (c. 800) on Java has 8,100 feet of relief carvings and over 500 life-sized images of the Buddha in nine rising tiers,

which become increasingly less realistic, more celestial and eso-
teric, until they end in formlessness.

But Buddhist art scarcely existed until the conversion of
the Mauryan monarch Ashoka (ruled c. 272–231 B.C.). Even
then, artists represented the Buddha symbolically as a tree, stupa,
or vacant throne, and human images of him did not appear for
another 150 years.

By then, of course, no one knew what the Buddha had
really looked like. Images of him varied at the outset, with two
schools arising at once. Alexander the Great had reached
Gandhara, in northwest India, and his influence led to Buddhas
wearing togas and Western mustaches. This style influenced
China; the earliest known Chinese Buddha is toga-clad. At the
same time further south, sculptors modeled Buddhas on slender
male fertility and prosperity symbols.

Like the Hindu gods, the Buddha needed marks of divinity. One technique was emphasis of the face. From the start,
artists often graced him with a halo or nimbus, like those of
Christ and the saints. It both drew attention to the face and conveyed radiance and charisma. Korean Buddhas had outsized
heads, which also highlighted the face. In Thailand and
Cambodia we see the Buddha sheltered by the hood of the huge
protective snake Muchalinda, which acts as a nimbus. Early
authors prescribed 32 other special insignia, such as a tuft of hair
sprouting from his forehead, webbed fingers, and lotus and
wheel marks on his soles.

But these strictures became harder to maintain as Buddhism
repeatedly forked. The Mahayana ("great path") Buddhists had a
more regal vision of him, and often decked him with a crown. In
Afghanistan, a Buddha carved in a cliff stands 172 feet tall, 19 feet
higher than the Statue of Liberty with its base. Even in nonage he
was impressive. Images of the baby Buddha taking his first step

showed him with arms upraised in victory, the gesture of a conquering hero. Eventually, Mahayana came to offer believers an afterlife in the Western Paradise, a heaven of wealth beyond mortal dreams.

The second main school, Theravada ("path of the elders," which Mahayanins scorn as Hinayana, "lesser path"), thrives today mainly in Sri Lanka and Southeast Asia, and emphasizes the lessons from the Buddha's earthly life. He is a role model, source of nirvana rather than postmortem riches, and Theravadin artists depict him more realistically, as a historical figure engaged in simple acts.

The Buddha is often a study in serenity, with a smile of transcendence that can approach inanity. Often too the Buddha is contemplative, as in the Seated Amida near Kamakura, where he folds his hands in his lap and looks down, his mouth slack with gravity. Buddhas can be skeletal, obese, thin-faced, heavy jowled, wide- or shut-eyed, smiling (often with hale-fellow grin), neutral, grave, jolly profound, thoughtful, mischievously hesitant. In some Thai statues he looks feminine, strolling along with a slinky, houri grace.

The variety ramifies in *bodhisattvas* ("enlightened beings"), rough equivalents of Christian saints. Avalokiteshvara had a thousand arms and eleven faces, often shown in a stack, one large head with smaller ones piled on top. By Tang China he had become the main figure in religious compositions, commonly shouldering the Buddha aside. Worshippers sought his image so eagerly that it spurred the development of block printing, which ultimately led to movable type.

The popular *bodhisattva* Maitreya, the Buddha of an incarnation yet to come, underwent a bizarre transformation after the Tang. He fused with a popular character called Budai, or Hotei in Japan, "hemp sack," a jolly fat man who wandered the world in quest of fun. He remains a favorite among Zen artists, who enjoy irreverence, and he is probably the source of the Western trinket Buddha — the potbellied buffoon with loopy grin.

In post-Tang China, too, another odd object of portrait arose: the *luohan* (*arhat* in India). These were ascetics, often of peculiar garb, who stood for individual achievement. They scowl or peer out with quizzically tilted necks, and their wild gaze made them favorite subjects of caricature. The *luohan* became a template for some Buddhas, and one from the Qing era shows him with a full Western beard, narrow face, large nose, and ears as long as bananas.

Monotheism rarely shows its God. He is one of a kind, with no need for visual identity. He stands apart from any incarnation and has traits that thwart the eye. "How can the invisible be depicted?" asks John of Damascus. "How does one picture the inconceivable? How can one draw what is limitless, immeasurable, infinite? How can a form be given to the formless? How does one paint the bodiless? How can you describe what is a mystery?"

Does the Christian God have a face? The Bible says he made people in his own image, which suggests he does. In the Old Testament, God appears in bodily form to Adam and Moses. Jacob says, "I have seen God face to face, and my life is preserved," and Isaiah saw him seated on a throne.

Yet actual depictions of God are rare. In the Greek Orthodox Church, the Great Council of Moscow (1666–1667) forbade images of God. In the West, he typically appears as a man with a long, flowing beard, as on the Sistine ceiling or in Titian's *Assumption of the Virgin* (1516–1518). In modern times, these images tend to the cartoonish, suggesting the artist isn't quite serious.

Why the ban? Theological concerns aside, depiction limits exaltation. It gives us an edge, shows us God's mood, individualizes him, gives him a sex, makes him one among many. Moreover, unless one adopts the Hindu solution of multiple limbs and heads, it suggests he is like us. Faceless, he is more powerful and mysterious.

Judaism also proscribes images of God, and long kept even his name a secret. But of the major creeds, Islam goes furthest.

Muslims forbid all facial images. When the Umayyad caliph 'Abd al-Malik placed his face on coins in the late 690s, the devout made such an uproar that he hastily replaced it with simple inscriptions. Ibrahim Pasha (c. 1493–1536) erected the only three statues in the 471-year history of the Ottoman Empire — of Hercules, Diana, and Apollo — and citizens cursed this potentate, though very quietly. Claude Lévi-Strauss blamed the emptiness of Islamic art on the rejection of images, which divorced artists from reality and bred a stale conventionality.

Yet the ban on portraiture is not in the Koran. It stems from custom and was never absolute. The Ottoman sultans regularly commissioned paintings of themselves, and the Mughals of India developed this art to a high pitch, especially in the reign of Jahangir (1605–1622). Jahangir even used portraits in diplomacy, and sent an artist to paint a likeness of Shah Abbas in Iran.

Even so, Islam is extreme. Yet sophisticates in most creeds agree the face of God is a human projection. It forges a link and helps people understand, even if it's just a dream.

Superportraits

The most bizarre faces stem not from literary imaginations like Mandeville's, but from visual ones. The pen makes the face infinitely plastic, and a cartoonist like Basil Wolverton can create galleries that make even fruit fly mutants seem banal. This free-dom to stretch and pull enables the caricature.

A caricature is a portrait in epigram. It exaggerates a few distinctive features, often for witty effect, and it has pinked the pompous for centuries. This facial parody arose with the ancient Egyptians, if not earlier. In the *Poetics*, Aristotle mentions one artist who improves faces, another who copies them, and a third who worsens them. They are the flatterer, the realist, and the caricaturist.

But caricature first reached eminence in the Renaissance, with printing. Church art was permanent, carved in stone or painted on walls. Printed art was fleeting. People could hide or junk it if the authorities came snooping by, and it brought greater freedom to criticize. Early moralistic cartoonists focused on the Dance of Death, since the afterlife bore special treats for the sinner. William Hogarth (1697–1764) became the first great caricaturist, etching *Gin Lane* (1751) and other images into the European psyche. Hogarth was basically a social satirist, though he did venture into politics, mocking Lord Bute, John Wilkes, and participants in the South Sea Bubble.

His British successors, Thomas Rowlandson (1756–1827) and James Gillray (1757–1815), took opposite paths. Rowlandson mocked social foibles, like the fashion-slavery of theatergoers. Gillray was probably the first full-time political cartoonist and he pie-faced Whigs and Tories alike. A crowd gathered outside his publisher's office whenever a new drawing from him was due. "It was a veritable madness," noted one observer. "You have to fight your way in with your fists."

Caricature has mass appeal, and Honoré Daumier (1808–1879) was its first great artist to exploit the lithography and rotary presses that made periodicals widely affordable. During the Revolution of 1830, he portrayed the politically eminent in the magazine *La Caricature*. "Every little meanness of spirit, every absurdity, every quirk of intellect, every vice of the heart can be clearly seen and read in these animalized faces," Baudelaire wrote. After new censorship laws and a stint in jail in 1832 for showing Louis Philippe as Gargantua, Daumier retreated to satire of manners. But all his work has aphoristic directness. As Baudelaire observed, "The central idea immediately leaps out at you. You have only to look to understand."

The best-known modern caricatures have appeared in political cartoons, by people like Thomas Nast, Herbert Block ("Herblock"), Pat Oliphant, Paul Conrad, Bill Mauldin, Jeff MacNelly. Nast (1840–1902) championed the Union in the Civil War — Lincoln called him "our best recruiting sergeant" — and invented the modern image of Santa Claus, the GOP elephant, and the Democratic donkey. He needled Tammany Hall relentlessly and one cartoon, entitled "The 'Brains,'" showed Boss Tweed with a bag of cash for a face. Caricature does not always delight its honorees. Richard Nixon resented Herblock's drawings of him in the *Washington Post* for decades, and Los Angeles mayor Sam Yorty once sued Paul Conrad for $2 million, in vain.

Not all caricature is weaponry. Zen Buddhists liked to portray Bodhidharma, the fifth-century founder of the Chan sect, in irreverent caricature, an unusual religious practice. David Levine's portraits in the *New York Review of Books*, while unflattering, are not normally malicious — though he sharpens his pen for special victims like Nixon and William F. Buckley. The faces of Al Hirschfeld even have impish charm. Yet all good caricature has tang. Some artists simply present scenes from life and warrant Baudelaire's cruel praise of Pigal: "He is an essentially *reasonable* caricaturist."

Caricature works by making us identify a real person with a comic image. We can look at a grinning, fantastically swollen face and still say, "That's Bill Clinton." The trick demands balance. On the one hand, the artist can't just exaggerate a single feature. Richard Nixon needs a distended nose, but also swollen jowls and a box chin, and intriguingly caricaturists generally agree on the key features to balloon. On the other hand, the artist must avoid excess. A cucumber nose and gross, saddlebag jowls would keep us from recognizing Nixon at all, and kill the effect.

Exaggerating the standout features may not be enough. Computers can do it, yet people typically draw more devastating caricatures. As in criminal identification, artists capture faces better than machines, so far.

Artist Annibale Carracci (1560–1609) said, "A good caricature, like every other work of art, is more true to life than reality itself," and psychologists now take this notion very seriously. Experiments show that we identify a good caricature faster than a complete and accurate line drawing. And surprisingly, caricatures also seem more realistic. In one recent study which compared genuine pictures and computer-generated caricatures, both of photographic quality, people picked the caricature as the better likeness. Distortion seems more accurate than truth. And familiarity actually strengthens the effect. With well-known faces, people named more extreme caricatures as the best likeness. It seems a paradox: The better we know a face, the worse we seem to see it.

Such findings have led some psychologists to call caricatures "superportraits." They suggest we recognize faces by using a shorthand of special features, which the caricature highlights and the true picture does not. The caricature feeds the vital variables straight into the formula, while the full line drawing adds clutter and gives all features equal weight.

Others go further. Psychologists Robert Mauro and Michael Kubovy have produced evidence that the brain stores faces as caricatures. Rather than a shorthand for recognition, the caricature lives intact in compressed memory. A cartoon of a grinning Jimmy Carter with teeth the size of playing cards actually matches the neural image better.

This theory has implications for the other-race bias. Mauro and Kubovy suggest that, in storing memories of other-race individuals, race may seem as distinctive as a long nose, and slot into the compressed image at the expense of more diagnostic features. Hence people can flounder at distinguishing members of that race.

Indeed, witnesses may have confused Laszlo Virag with Georges Payen partly because, though their faces clearly differed, they shared distinctive features caricature would enhance, like thin lips and pointed nose.

"No everyday experience is too base for the thinking man," said Umberto Eco, and the surprises beneath the lid of the caricature bear him out.

A Fount of Identities

Masks are toys of self. They are donnable faces, instant personas, the most immediate and widespread kind of disguise. From kachina ceremonies to the Feast of Fools to stagecoach robberies in the American West, they have let people play with identity. They have made us into criminals, paramours, and gods.

Like many other cultural innovations, masks date back to the Upper Paleolithic. A skeleton buried 27,000 years ago at Dolní Vestonice in the Czech Republic may have worn a painted mask. It lies with two other teenagers in a possible love triangle, since one has a hand on another's pubic area. But a limestone mask from Hebron, dated to around 6500 B.C., is the earliest we can be certain of. Stone masks are uncomfortable but they last, and frailer kinds surely preceded it.

Skulls make potent masks. They let a former face lie atop a current one, bone to living skin. They are reality, not representation, and hence are powerful ties to ancestor-gods. Yet a skull has wide holes and may need prosthesis. For instance, archeologists have found twelve likely skull masks at Jericho dating from between 7600 and 6000 B.C. Artisans had reshaped them with plaster of Paris and painted at least five. Melanesians, Aztecs, and Nigerians all made skull masks. The Melanesians resurfaced the cranium with wax—often filling in eyes, sculpting noses, and gluing on human hair—and painted it. An Aztec skull mask of the scrying god Tezcatlipoca in the British Museum has eyes of pyrite and strips of turquoise mosaic and obsidian across its gaping face. Montezuma may have presented this object to Cortez.

The skin of dead humans has been mask material. The Mapuche of Chile dried the facial skin of slain captives and wore it at dances. In Mexico, among the Huastecs and later the Aztecs, the most important masked god was Xipe Totec, "Our Lord the Flayed One," the god of springtime. Young Aztec men imitated him, covering their faces and bodies with a mask of skin from a human sacrifice. They wore it for twenty-one days, by which time it rotted and a renewed person emerged.

Statues have worn masks. When an Aztec king fell ill, priests placed masks on idols of the gods until he recovered. They did the same after a public disaster. Statues can also *be* masks. One anthropologist saw a sculpture in northern Ekiti that was over six feet high and weighed over 120 pounds, yet was an *epa* mask, meant to be worn.

No one knows why masks first arose. They may have begun in hunting, as aids in approaching prey. In Nayarit, Mexico, people once wore Cora masks in hunting rituals, and today, during

Holy Week celebrations, men don Cora masks and track and slay Jesus as if he were a deer.

They may have begun as tools of otherworldly power. Masks were the first idols. They render gods and spirits well, partly because they bestow nonhuman faces. For instance, Egyptian priests may have assumed the role of their animal-headed deities in masks. Moreover, many primitive peoples believed spirits inhabited the head, and from there it was a short step to externalizing them with a mask.

Masks have long represented the dead. Death masks preserve likeness and can become icons of a new ancestor-god. In New Ireland, people gathered in June to mourn the dead. Under the aegis of a secret society, they wore masks of the departed, and as the crowd recognized the dead person, it cried out his name and made lamentation.

Masks can confer ghosthood. Masked young men garbed as ghosts commonly made juju — magic — as they wandered the villages of upper Nigeria, Dahomey, and Togo receiving gifts. Among West African tribes like the Ibo, a brief resurrection occurs. The spirit of the deceased appears aboveground twice a day after burial, wearing a pale mask and speaking through a kazoo-like instrument. People carefully brush the grave dust off and lend it a shoulder to lean on. Over four or five days, the spirit slowly gains strength until it can walk about unaided, at which time a second, ritual burial commences.

Masks can be the homes of gods. At one time the deified ancestors of the Hopi, called kachinas, dwelt on earth between the winter and summer solstice. But when they could remain no longer, they taught the Hopi to make masks for their spirits to inhabit. The Hopi first don these masks at a late-December ceremony called The Return of the Kachinas, and retire them at the end of June.

A mask can even *be* a god. The Delaware Indians have a god called Living Solid Face, a mask that is alive and divine. Living Solid Face taught the tribe to make masks in his likeness, and promised he would perfuse them when worn. He acts as a generally benign guide for the tribe and its hunt.

This link to the supernatural has myriad facets, and indeed a mask is the face of dreams. With it we can be anyone we can imagine. Thus a medicine man in the right mask could cure a specific ailment. In Sri Lanka, there was a different mask for each of nineteen diseases. The shaman donned the mask, danced in front of the patient, and lured the spirit out of the victim and into himself. He then walked to the edge of town and pretended to die, ridding both himself and the town of the pest.

Masks can clean the environment. According to explorer K. T. Preuss, the Kagaba of northern Colombia believed masks were originally the faces of demons, who had given them to humans so they could perform the dances that ward off illness. Over time, the sexual sins of villagers slowly crumbled local magic stones into a powder which fostered disease. Since the sun was a god who served humanity, the great sun-mask swept this dust to the ends of the earth.

Masks, like totem poles, can mock. At potlatches, British Columbian Indians sometimes wore masks of rival chiefs. Supposedly, the heat of flaming property would burn up half the face of these masks. Masks could also mock by foisting identity on the unwilling. Protestant women accused of witchcraft had to wear "masks of shame" at their trials. These often had long noses, wild grins, huge ears, and frazzled hair, and were a punishment before the verdict.

The purposes of masks roll on and on, like the desires of people themselves. The Dan of the Ivory Coast use their most important masks to communicate with the lofty god Zlan, by way of ancestors. But they also have sacrificial masks, on which they make offerings to forebears; avenging masks, to aid police and judges; initiation masks, to teach novices and regale the village; *sagbwe* masks, to guard towns near the forest from threats like fire; and, at the lowest level, entertainment masks.

And masks have played multifarious roles in the cultures of India, China, Japan, black Africa, ancient Egypt, pre-Columbian America, and Maori New Zealand. People have donned them to accent ritual dances, to frighten the enemy in battle, to appease the gods, to accompany a ruler to the afterlife, and to commune with divinity.

Masks need not even represent individuals. They can depict societies and geophysical events. The towering *ijele* masks of the Nigerian Ibo, nine feet wide by twelve to eighteen feet high, show Brueghel-like tableaux of the daily life of people, animals, and deities. In New Britain, the *mandas* ceremony portrays events of mythic times with 84 masks. One shows whirlpools and stands for the birth of the ocean, and another depicts the parting of the earth from the sea. Among the Senoufo of Ivory Coast, Mali, and Burkina Faso, the *kponiougo* mask recalls the original chaos of the world, before Koulo Tiolo imposed order on it. The *mosh'ambooymushall* mask of the Kuba of Zaire shows the beginning of the world.

Today in the West, masks more often blot identity than transform it. A child on Halloween dresses like a demon or a cowboy, but doesn't act like one, and masquerade-goers adopt conversation-piece identities. The face no longer betrays them and the fetters of identity fall away. Hence many feel freer, less accountable, and their secret selves bloom. A mask can sometimes reveal a person.

Veils and fans are semimasks. They hide discreetly and par-
tially, and thus enhance beauty and aid dalliance. In Victorian
and Edwardian times, women often wore veils to call attention
to the face, yet to gauze it in mystery. They held the face apart,
almost aloof from the world. These veils, often of very wide
mesh, also dimmed wrinkles and other blemishes, and could
make the facial features dreamily vague. And by setting up a
boundary, even a near-transparent one, between the face and
the observer, they created a showcase.

At the end of the 1600s the fan became a popular way to
coyly hide and reveal the face. Reynolds and Fragonard painted
fans and a special code soon evolved around them. For instance,
the number of waves of a half-opened fan signaled the time for
an assignation. A fan touched to the lips indicated the woman
wanted a kiss. A fan pressed to the heart showed by its intensity
the strength of her passion. A Spanish publisher issued a fan dic-
tionary to explain the lexicon, and indeed fans grew popular in
Spain just when custom began to limit conversation between
the sexes.

Real masks are an open sesame to all kinds of semilicit
frolicking. During the Roman festival of Saturnalia, they helped
upend the social order. Serfs masked as nobles, nobles cavorted
ignobly, and officials smiled on the ensuing crimes. In the
medieval Feast of Fools, peasants laughed and drank in masks,
and priests turned their vestments inside out, donned bishops'
masks, and staged the rites of the Lord of Misrule, the Pope of
Unreason. In 1207 Innocent III forbade masks, to little effect,
and when papal authorities finally drove them out of the church
in the 1400s, they moved over to Carnival.

Carnival is the annual apocalypse gala before the death-time
of Lent. At the Venetian Carnivals of the seventeenth and eighteenth centuries, men and women wore masks on the piazzas and

in candlelit ballrooms, and the anonymity emboldened them. "Old and young, patricians and plebeians, rich and poor, all are disguised," wrote Pompeo Molmenti (1852–1928). "A harlequin murmurs sweet nothings into the ear of a young woman in a domino who laughs and takes refuge among the crowds. A mattacino, in white with red garters and red shoes, throws eggshells with rosewater at patricians' windows." Venetians have recently revived Carnival, and it is a mob of masked tourists.

Masks have always cloaked sexual daring. In Shakespeare, Romeo moves among the Capulets in a mask, and Viola woos Olivia in one. The eye mask, a strip of cloth across the eyes, was popular in Venice and ideal for the purpose. It freed the lower face for speech, and hiding the eyes slows recognition more than any other part of the face. Even so, it revealed enough that men and women could gauge each other's youth and attractiveness.

Masked balls became popular in America in the mid-nineteenth century, and the tradition continues, though sometimes in a different key. Mardi Gras revelers may wear masks that change sexual or ethnic identities. And masquerades, especially on Halloween, can be the province of prostitutes, fetishists, and others with sexual secrets. When masks are genuine disguises, they show where society draws the line.

European bandits did not begin to wear masks until the 1500s. In addition to hiding the face, they often served as a fashion statement, or at least a way to tell highwaymen from the hoi polloi. Many of these criminals came from the upper classes, and needed cash to pay gambling debts, make up for lost inheritances, or regain wealth lost in war. A gentleman in particular needed a mask, since his victims today might be his dining companions tomorrow.

Highwaymen wore a variety of disguises. The sixteenth-century bandit Gamaliel Ratsey, a squire's son, donned a hideous mask to unnerve his victims. Jack Collet dressed as a bishop and rode with four or five companions garbed as servants. Tom Royland and Thomas Simpson posed as women. Mary Firth, or Moll Cutpurse, disguised herself as a man and grew fairly wealthy. Francis Jackson, a gambler, gallant, and bon vivant, turned highwayman to support a prostitute girlfriend. He became an expert on disguises, and astutely recommended owning many wigs and beards of different colors. The creation of a highway patrol in the early nineteenth century ended this profession.

In the nineteenth-century American West, bandits pulled bandannas up over their noses, a trick learned from miners, who used them to keep dust out of their throats. Black Bart, who held up Wells Fargo stagecoaches in California between 1875 and 1883, wore a derby over a flour sack with eyeholes. He was always a gentleman, but made a career-ending mistake when he dropped a handkerchief at a robbery site. Detectives traced the laundry mark to San Francisco and found him to be Charles Bolton, a small, balding man who favored derby hats and posed as a wealthy mine owner. After his arrest, citizens besieged the local prison to see the famous personality.

The mask is a business necessity for the robber of today, given surveillance cameras, ID technology, and the power of TV to spread his image across the globe. Criminals generally wear easily bought items, such as Halloween masks, ski masks, and nylon stockings (which, incidentally, work not by hiding facial features, but by flattening them).

Executioners wore masks that covered most of the face, to prevent public identification and stigma. No one knows who beheaded Charles I. The hangman in Ireland wore a hideous mask, and also a wooden hump on his back to deflect stones flung at him, since he was often executing popular rebels.

But as executions became spectacle, with London merry-makers lining the streets from Newgate Prison to Tyburn and *le beau monde* renting balconies overlooking the route, the executioner removed his mask and became a public figure, almost a revelmeister. At the same time, the condemned began putting masks on, often mere blindfolds or handkerchiefs. Sir Walter Raleigh scorned the mask, saying, "Think you I fear the shadow of the axe, when I fear not the axe itself?"

All such masks protected reputation. And sometimes they could make it soar.

Unknown Faces

On November 19, 1703, a mysterious man died in the Bastille. He had arrived there in 1698 from the prison at Pignerol, where he'd been since 1661 — or 1669 — and he lived in well-guarded isolation under the custody of the most trusted jailer of Louis XIV. Officials buried him under the name "Marchioly." History might have forgotten him, but he teased it intolerably by constantly wearing a black velvet mask.

Voltaire was the first to publicize him. The witty author, who for some reason claimed the velvet mask was actually iron, had himself dwelt in the Bastille a few years later and seemed to possess authority on its affairs. In his *Siècle de Louis XIV* he said the inmate was plainly an important man and that no such figure had disappeared at the time his captivity began. He left readers to infer the Man in the Iron Mask was an illegitimate sibling of Louis XIV.

The prisoner's identity soon became a public guessing game. Some said he was a twin of Louis XIV, and Alexandre

Dumas (1802–1870) has probably immortalized the notion in *The Man in the Iron Mask* (1848–1850). Others advanced a fantastic series of candidates. He was a French admiral who vanished in Crete. He was Molière. He was a British lord, an Italian intriguer, a daughter of Louis XIII, the head of a vast criminal conspiracy, and in a theory offered as recently as 1978, the supposed leader of the supposed resurrection of the Templars. According to one saying, the Man in the Iron Mask has had more incarnations than a Hindu god. Who was he? The best modern guess is Eugène Dauger, son of a lady-in-waiting at the court. Why did he wear the mask? No one knows.

But the most interesting Iron Mask question is: Why should anyone care? Why has interest in this shadowy and insignificant figure persisted over almost two centuries?

At root, of course, researchers yearn to yank the mask. They are like Christine Daaé in *The Phantom of the Opera*, who tries to snatch off Erik's disguise as he rows her across a subterranean lake. He warns, "You are in no danger, so long as you do not touch the mask." But the impulse is irresistible. "Suddenly, I felt a need to see beneath the mask," she says later. "I wanted to know the *face* of the voice, and, with a movement which I was utterly unable to control, swiftly my fingers tore away the mask. Oh, horror, horror, horror!"

Daaé is an extreme case, but everyone has felt the urge. We need to know people's faces. We want a photo of the author on a book, and once we see it, we're satisfied. It scarcely matters what the author looks like.

This basic face-hunger has enriched shrewd entrepreneurs. In 1933 Detroit radio-station owner George Trendle decided to create a Western around a Robin Hood character with a mask, whom his scriptwriter Fran Striker fleshed out: the Lone Ranger. He wore a mask, according to an early explanation, because brigands had once slaughtered a group of Texas

Rangers, of whom he alone survived. If gang members knew he was alive, they'd pursue him. This fearful rationale later evolved into a heroic blurring of identity. Masked, the Lone Ranger could be any one of the slain lawmen, and stood for all of them.

As the Lone Ranger rocketed into Americana, the device found imitators. Batman appeared in 1939. One night when he was a child, robbers slew his millionaire parents as they strolled home from a movie. The youngster dedicated himself to preventing such crimes in the future and eventually donned a bat disguise to frighten superstitious criminals. A gallery of superheroes — Spiderman, Captain Marvel, Zorro, Captain America — also came to wear masks.

The explanations for these masks sound so odd and weak because their real purpose can't be uttered. It is manipulation. The mask breeds mystique. It creates a secret, and as sociologist Georg Simmel (1858–1918) noted, secrets adorn and enhance personality. They cache information, and the face is a stream of it. Simmel believed that, to the ordinary person, superior individuals always had an aura of mystery, and a mask bestows one instantly. Indeed, the contract for Clayton Moore, the actor who played the Lone Ranger, called for him either to appear in public masked or to remain anonymous. Hence the role became an icon instead of the actor.

Masks also spur dramatic questions: What does he look like? Who is he? Viewers wanted to know the Lone Ranger's face, and so did his enemies, who regularly threatened to rip off his mask once they captured him. They sought to demote icon to individual, as the Mexican government attempted in February 1995, when it revealed the birth name of guerrilla Subcomandante Marcos. In the case of the Lone Ranger, this threat could create a moment of exquisite tension.

★   ★   ★

Hundreds of portraits of Joan of Arc (1412–1431) have come down to us, and her image graced millions of French homes long before the Church canonized her in 1920. Most showed her either as a pious shepherd or a warrior atop a rearing steed, and they depict her face with great specificity. They are all fantasies. The Maid of Orleans never sat for a portrait and we have no idea what she looked like.

Confronted with true unknown faces, we tend to fill in. We conjure faces of people we haven't met, and everyone has heard the line: "You don't look anything like I expected!" For years Honoré de Balzac allowed no one to paint his picture. When he finally relented at thirty-seven, the portrait showed a man with bulbous nose and double chin. Sighed a women's magazine, "Ah! the illusions that will be lost at the sight of that painting! The voices that will cry, 'Give us back our Balzac!'" Women had been filling in.

An unknown face can spur quests. No known portrait of Thomas Pynchon exists since he studied at Cornell in the fifties, and journalists occasionally try to hunt him down. In *Dr. Jekyll and Mr. Hyde* (1886), Hyde is curiously faceless. Witnesses see his countenance but can't describe anything about it, except its incredible evil. Hence the lawyer Utterson feels "a singularly strong, almost an inordinate, curiosity to behold the features of the real Mr. Hyde," and begins his search for him.

Artists gladly satisfy such cravings. For instance, Rembrandt's *Aristotle Contemplating a Bust of Homer* (1653) shows portraits of Homer, of whom we know nothing, and Aristotle, of whom we have no reliable description and who wears the garb of a Dutch master. The great artist simply invented them.

Artists in antiquity tossed off such pseudolikenesses with disturbing ease, leaving classical art studies with a major problem. They often made portraits hundreds of years after their famous subjects had died. The images we possess of Aeschylus, Sophocles, and Euripides seem posthumous concoctions, and the best-known sculptures of Aristotle and Demosthenes clearly are.

Such works could look highly individual, since the artists often modeled them on friends. A hanger-on goes down in history as Plato. For instance, in Rubens's *Four Philosophers with a Bust of Seneca* (1611–1612), a quartet of men sits talking beneath a marble bust of the Stoic. Rubens drew the head from another work of art, and we know today it is not Seneca, but some unidentified person from antiquity whose face lives on through imposture.

Some ancients complained of the state of portraiture. Pliny the Elder (23–79 A.D.) mourned that likeness had gone out of fashion. "Heads of statues are interchangeable," he said, adding, "sarcastic verses have gone the rounds on this topic." Yet people "cram the walls of their galleries with old pictures and revere portraits of strangers." Today, no bust, portrait, or other representation of Pliny survives, nor does any physical description.

When his students asked the philosopher Plotinus (205?–270 A.D.) to have his portrait done, he questioned the enterprise. His face wasn't him, he said, but merely his husk, and a portrait would therefore be a husk of a husk, an illusion doubled. His students demurred, of course, since for them Plotinus was very much in his face. They eventually did convince him to sit, but the portrait is lost and we don't know what he looked like. Plotinus has prevailed.

We have no contemporary portrait of Christopher Columbus, but we do have descriptions. He was long-faced, with an aquiline nose and clear eyes. Observers variously call his hair red, blond, and white. Yet at the Chicago Exposition of 1893, visitors saw 71 alleged originals, all varying wildly. Some showed his face lean and long-jawed, others puffed with fat. He was clean-shaven, mustached, bearded. In some he resembled the aged Rembrandt and in others the young Olivier.

We have no image or description of Moses or the Buddha. Millions of pictures of Jesus exist; none is reliable. Early ones show him beardless and some show him blond. We have a whole village of portraits of Alexander the Great; scholars don't know which, if any, is a likeness. We lack even a wisp of information about Lao-tzu, who may be a historical phantom. We have no description of buccaneer Henry Morgan, except near death when rum and tropical disease had ravaged his face, nor of the famous pirate called Mrs. Cheng, who ruled the South China Sea in the early 1800s with her 17,000 men and 224 junks. Such faces have simply disappeared, but we supply them anyway, just as we paste a face on God.

In *Darkness at Noon*, the jailed Rubashov begins tapping messages to the man in the next cell. As he responds, Rubashov imagines his face. First he thinks him a despairing innocent with a black Pushkin beard, his face unwashed and slovenly. After a sharp retort, he deems him a conformist, clean-shaven and fanatical. Then he concludes the man is a royalist, and sees him as a young officer, handsome and dumb. Slowly this individual develops a monocle and a mustache, curled at the ends. Rubashov is talking to a total mask, and facial images well up in him.

The Internet can now make a Rubashov's cell out of any room. We can tap a keyboard and talk with unseen people all over the globe. The phenomenon is more immediate than e-mail, which itself has so proliferated that characters in Douglas Coupland's *Microserfs* (1995) call normal socializing "FaceTime." In blind conversation parlors like The Well, people routinely have secret identities, complete with "pseuds." The normal moorings to self—face, voice, body, name, gender, race, history—vanish.

Beyond the pseud, ID is one's verbal style, often an invented persona. Reciprocal knowledge can thin to near-vapor. It is prosopagnosic heaven.

Faceless intimacy has also arrived. Phone sex has become a cultural set-piece, almost a cliché. Nicholson Baker's *Vox* (1992) is a long phone-sex causerie between two utter strangers. In the movie *Denise Calls Up* (1996), busy individuals interact solely by phone, and two characters cycle through a sexual relationship start-to-finish without even seeing each other's faces. In *The Truth About Cats and Dogs* (1996), a first sexual encounter occurs over the wires, and the plot hinges on a man's mistake about a woman's face.

Yet even on the phone, voice identifies us. Online, in the forums called TinyMUDs, people can select gender, from up to ten choices, and even species, and invent age, job, history, and selves. They chat or engage in ritual combat. They may also have tinysex, the keyboard version of phone sex. It's a verbal masked ball with assignation rooms, faceless intimacy to an unprecedented degree. TinyMUDs show how normal cues to identity keep us honest. In these worlds deceit is commonplace, in gender, orientation, looks, and past. Men notoriously pose as women, and less often vice versa. People wipe out their histories with a change of pseud. It is the ultimate masquerade.

Of course, cyberchat and TinyMUDs are voluntary and recreational, adventures in identity rather than identity itself. In a world of total cyberchat, identity Flatland, we'd have no idea whom to trust, or whom we were really talking to. Beyond a base of shared knowledge or devices like coded passwords, we couldn't recognize business partners or even spouses.

Some online services like The Well routinely hold parties where people can meet in FaceTime. Ultimately, even the best keyboard talk can't replace the face.

III Semaphore
Semaphore

The face is the soul of the body.

LUDWIG WITTGENSTEIN

Four
4

The Skin Code

N late 1897 Harry Houdini and his wife Bess, desperate for cash, joined a traveling medicine show called Dr. Hill's California Concert Company. Its members typically arrived in a small town, planted themselves on a street corner, and brought out musical instruments. Houdini whacked a tambourine and Bess sang. When enough people gathered, Dr. Hill mounted a platform, extolled his splendid elixir, and made sales. He also announced a performance in the local hall that night.

There, Houdini and Bess did a mind-reading routine, posing as spirit mediums. Bess walked out into the audience, took a dollar bill from a spectator, and asked Houdini to divine its serial number. Houdini concentrated, looking intent and unearthly, then slowly reeled it off, to the general murmur.

The trick had an interesting secret. They used a code with a giveaway word for each digit:

1 = Pray	6 = Please
2 = Answer	7 = Speak
3 = Say	8 = Quickly
4 = Now	9 = Look
5 = Tell	0 = Be quick

If the serial number was 48655971677, Bess might call out: "Now, mind reader. Quickly, please! Tell us what you see! Tell us! Look into your heart! Speak, I pray you! Please, speak to us! Speak!" And the enlightened magician would oblige.

We all do genuine feats of mind-reading every day. The trick involves the face, and we perform it much as Houdini read serial numbers. For faces too have a special code. Its basic elements are about as few as Houdini's, though its shadings and fusions are myriad. They are facial expressions.

Facial expressions are ancient and have had eons to enter the DNA. Reptiles have facial signals, such as opening the mouth, with or without bared teeth. Fur seals and walruses greet, threaten, and submit with their faces. We can read a dog's countenance, to some degree, and monkey faces grew hairless to improve signaling. We humans laugh, weep, blush, and show an array of subtler expressions whose meanings time has imprinted on our brains.

As a result, most of us can't explain even their fundamentals. It is as if a little Houdini dwelt within each of us, deciphering expressions in a hermetic cell and passing on the results. We seem to view not muscle movements, but the shifts and surges of the mind itself.

These expressions often transcend words, since they predate them. We've all seen looks that blossom into meaning beyond description. Jane Eyre mentions a "vague something" that occasionally opened in Rochester's eye which made her "fear and shrink, as if I had been wandering amongst volcanic-looking hills, and had suddenly felt the ground quiver, and seen it gape." The Japanese have two words roughly equivalent to "soul": *ki* and *kokoro*. *Ki* is the part that responds almost automatically to the world and *kokoro* is the secret part of the heart. The face shows *ki*, and sometimes *kokoro* as well.

For millennia this code has eluded scrutiny, and few aside from gifted artists and novelists have grasped it. In the last thirty years scientists have partly deciphered it. They have found six basic, universal expressions and a handful of other candidates. They have also explored the subtle signals called paralanguage, the even subtler ones of the eye, and the mysteries of weeping, smiles, laughter, blushing, and staring. It has all wrought a revolution in our knowledge of the face.

But before it, there was physiognomy.

Low Brows and Lowbrows

Everyone's face is candid and cryptic by turns. It is like a treasure map of character, full of invisible and shifting lines, a boon for its decoder. And for centuries seers have arisen to fill this role. They have practiced physiognomy, the divining of character from facial contour.

The profession is old as the redwoods. Face-readers plied their trade in China before Confucius, and Aristotle devoted six chapters to physiognomy in *Historia Animalium*. Polemon (2nd cen. A.D.) brought respectability to it as the Roman Empire was edging past ripeness, and left a tribe of professional face-readers in his wake. Avicenna (980–1037) discussed it seriously, as did Albertus Magnus (1193?–1280) and the Persian scholars Ali ben Ragel and Rhazes.

Michael Scot, astrologer to Frederick II, wrote the first printed book on the topic, *De Hominis Physiognomia*, penned in 1272 and published in 1477. This work led to a weird five-hundred-year entanglement of physiognomy and astrology, in which experts read destinies in the lines of Jupiter, Venus, and Mercury on the forehead.

The modern prophet of the craft was Swiss pastor and poet Johann Caspar Lavater (1741–1801), who claimed he could divine character from a black silhouette. He loved oratory and spoke fervidly of physiognomy as a benefactor of humanity. Goethe became a friend and traveling companion, and Lavater's four-volume opus *Fragments of Physiognomy for the Increase of Knowledge and Love of Mankind* (1775–1778) made him famous. He treated his talent as an inborn knack, beyond codification. "What folly it is to make physiognomy into a science," he declared, "so that one can speak on it or write about it, hold seminars or listen to them!"

Lavater sparked a face-reading craze in late eighteenth-century Europe, and it persisted into the nineteenth, imbuing the work of major writers. When Jane Eyre meets Rochester, she gives his face a quick once-over. His nose is decisive, she thinks, and his full nostrils indicate choler. Balzac peppered his novels with such analysis and in 1839 wrote a staunch defense of accused murderer Sébastien Peytel. Peytel's guilt now seems clear, but Balzac believed his claims of innocence because his facial contours ruled out hypocrisy.

Cesare Lombroso (1836–1909), a nineteenth-century "psychiatrist," devised a physiognomic system for identifying criminals. The born malefactor, he said, had prominent rodent-like incisors, either a receding or large and flat chin, little or no beard, premature wrinkles, and bushy eyebrows which either met or rose devilishly at the outer ends. He declared that criminals rarely go bald, or even gray. Murderers more often have dark hair than blond, and swindlers tend toward curly hair. Of course, the list of Death Row denizens with small teeth, normal chin, beard, smooth skin, or balding pate is a long one, and today Lombroso inhabits the bad-boy corner of science, along with Velikovsky, Burt, and Lysenko.

Franz Joseph Gall (1758–1828) followed up on Lavater with phrenology. While physiognomists examined face structure for character, phrenologists scrutinized bumps on the head for memory, reason, and imagination. Gall disliked physiognomy, but the two techniques meshed and many practitioners employed both. Phrenology was a learnable system, not a gift from beyond, and it inspired a few physiognomists to set down precepts for their own field, which made it seem more of a discipline.

Today its vogue has passed, perhaps because of the river of evidence we see on TV. Yet physiognomy remains with us. A 1993 book from a reputable American publisher falsely suggests Darwin smiled on the practice and informs us that mastery of face-reading can give us a competitive edge in today's fast-paced business world.

It proffers a range of entertaining and often startling information. Interviewers study the facial contours of job applicants, we learn, seeking a protruding chin ("dominance and lust for glory"), a deep philtrum ("determination"), and rectangular eyes ("shrewdness"). A mole below the middle of the lips indicates a serious thinker, and one on the bridge of the nose shows tremendous sexual desire. A large nose tip suggests violence. A cleft chin betokens conceit, love of publicity, often hypocrisy.

The book also liberates physiognomy from the repression of the past, revealing the most intimate secrets to a casual glance. People with thick, crescent-shaped eyebrows or a concave mid-nose experience "the most fantastic orgasms known to mankind. . . . [They] are among the luckiest people in the world." A man with a large nose and broad nostrils has a "strong penis, but lacks stamina." A short-chinned woman with broad nostrils has a "deep vagina; not easily satisfied."

Is physiognomy valid? It posits genes that link facial surface to mental and emotional traits. In fact, no a priori fiat excludes such bonds. The genetic code yields many odd associations. For instance, a diagonal cleft in the earlobe comes with greater risk of heart attack.

But scientists deem it taradiddle. Darwin called it "the so-called science of physiognomy" and said whatever truth it had depended on different people using different facial muscles more, hence increasing certain lines in the face. Psychologist Thomas R. Alley says, "Physiognomy is, with few and nearly negligible exceptions, an invalid practice." It confuses structure with mind, the resting face with its dance of signals.

Physiognomy can damage anyone foolish enough to take it seriously. Captain Robert Fitzroy almost rejected Darwin as naturalist on the *Beagle*, because, as Darwin wrote in his *Autobiography*, he "doubted whether anyone with my nose could possess sufficient energy and determination for the voyage." It misled Balzac about the murderer Peytel, and it would blind anyone relying on it in hiring or romance.

Yet a folk physiognomy persists. People utterly ignorant of Lavater make judgments from facial stereotypes, often predictably across cultures. As scientists point out, the studies in this area may not be entirely reliable. Many use photos or drawings, which are two-dimensional and static, though faces themselves are normally in motion. Expression subtly alters our sense of contour, and it's often hard to tell which features people are really responding to. But the basic results remain.

For instance, individuals in many cultures rate faces similarly on dominance. They link it to signs of age — receding hairline, broad face, smaller eyes — and some psychologists believe these features evolved to show seniority. Subjects also associate dominance with traits like thin lips and low eyebrows, which suggest anger. And they view a prominent chin as a mark of will and ambition, while "weak chin" has entered the general lexicon.

Some features convey warmth. Large, Bambi-like eyes seem kinder. Naturally high eyebrows correlate with affection and submissiveness. Studies have found we deem the babyfaced warmer, weaker, and more honest than the mature-faced. Hence advertisers cast the babyfaced where credibility depends on trustworthiness, the mature-faced where it hinges on expertise.

We like smiles. They increase assessments of intelligence, humor, kindness, and honesty, and we also bestow these qualities on people with naturally upturned mouths.

People use facial contour as a guide to intelligence. They view subjects with large, high foreheads as bright, and the facially deformed as criminal or stupid. In fact, face structure bears virtually no relation to intelligence.

The lure of physiognomy, of course, is theory of mind, which we are attuned to seek out. The promise is an easy route to it. Like astrology and auras, physiognomy can give us rapid, secret knowledge of other people's character, let us slip past the obstacles set up by evolution and described by game theory. It's a kind of magic, freighted with the exhilaration of satisfying our endless, impossible need to know.

Soul on Canvas

When Roger Fry saw Sargent's portrait of General Sir Ian Hamilton, he said, "I cannot see the man for his likeness."

Likeness is surface truth and can be almost a blank. Good portraits take us into a face, present a mind made flesh, and they can fascinate us almost like living people. No painting is more famous than the *Mona Lisa*, and Botticelli's *Venus* and Michelangelo's *David* dominate the list of best-known artworks on earth.

A mystique has arisen around the portrait. "A man's whole life may be a lie to himself and others," wrote William Hazlitt (1778–1830), himself a former portraitist, "and yet a picture painted of him by a great artist would probably stamp his true character on the canvas, and betray the secret to posterity." In *The Picture of Dorian Gray* (1890), the portrait takes on magic powers and changes along with its subject's soul. It is more revealing than his face.

How do artists stamp the canvas with a soul? In the painting of Gray, character reveals itself partly through physiognomy, and the image grows uglier as Gray becomes more corrupt. Genuine artists have also resorted to physiognomy, so for instance the intellectual sprouts a higher brow. Such portraiture enjoyed vogue in the ancient Roman Empire, the Renaissance, and the eighteenth and early nineteenth centuries when Lavater was popular. Indeed, some modern art critics still trot out physiognomy in the scrying game of portrait interpretation.

But expression is the normal key, and artists have approached it from many angles.

One tack is the summation. Painters have sought to fuse many moments into one. They shied away from strong emotions, since by freezing a second of surface gust they might miss the sitter's depths. In *Philosophy of Art* (c. 1802) Friedrich Schelling said the portrait must collect the many instants of a person's life into a single image which is "more like the person himself, that is, the idea of the person, than he himself is in any one of the individual moments." But the peril is obvious: How well can an expression reveal a sitter if the sitter has never shown it?

Summation, along with sitter vanity and a yen to address the ages, tends to give portraits a composed formality. The sitter stares into space, hushed and marmoreal, a face for eternity. This stasis partly reflected long hours of posing and eased work for the artist, of course, but it entombed countenance.

The Expressionists shredded and stomped composure. Their canvases squirm with agony, most notoriously in Edvard Munch's *The Cry* (1893), in which the world shudders like a gong. They and other painters sought impression and often chucked likeness, sensing like Plotinus that it limits direct knowledge of the soul. In his late portraits Van Gogh shunned close study of the model and painted his sense of the hidden personality, trying to shape "apparitions" for a future age. Oskar Kokoschka said that in his early work he sought an incandescent essence, like the memory image that is more vivid than the flesh-and-blood person "because it's concentrated, as if by a lens."

Yet the quest for such an essence led through the artist's mind, often a singular journey, and in the end the painter's character could merge with the sitter's, or even replace it. For instance, Francis Bacon's portraits all look hazed or violently smeared because his own anxiety suffuses them. The hunt for the subject becomes a hunt for oneself.

"We simply cannot check our interpretation of the other, our construction of his inner nature," wrote Simmel. The best portraits show a single, evocative expression and lock us into this mind-reading, much as rustling paper on the floor makes a cat stalk and pounce.

Yet capturing a precise look is "notoriously difficult," as art critic E. H. Gombrich observed. One German bank official shrewdly insisted that all currency bear a portrait, because expression is the hardest thing for a forger to imitate and hence the best flag of a fake. Gombrich felt the principle applied to art forgeries as well. Contemporary artists can't help but give faces a "modern" expression, which experts find "easy to spot but extremely hard to analyze."

Artists like Rembrandt developed a posthumous reputation as possessing a vast memory of facial expressions, though such an inner catalog may not be necessary. Rodolphe Töpffer (1799–1846), inventor of the cartoon strip, noted that a student of expression need only draw a face on paper. That face will of necessity have some expression. And to find its secret, the artist can play with the facial features, extending the smile, lowering the brow, tightening the lips. Da Vinci and Hogarth made many such trial-and-error experiments and they yielded a vast array of expressions.

Some artists like Hals, David, and Ingres specialized in catching the flash of life, the secret or delicious glance, the flare of excitement, and their approach grew more common after the inroads of the camera. Manet excelled in awkward or off-guard instants, as in *A Bar at the Folies-Bergère* (1881), and Edgar Degas almost made a genre of them.

Others render both persona and the soul beneath, face and subface and their quivering interface. For instance, Raphael captured the mild yet ambitious visage of Baldassare Castiglione, author of that symposium on good taste called *The Courtier* (1528). Castiglione's controlled face suggests a cautious climber, a student of the right steps. He taught patina and, ironically, his surface here betrays him.

At the highest level, artists show "the movements of the mind," as Fréart de Chambray called it in 1662, and da Vinci's *Mona Lisa* (1503–1505) is a classic example. It is a portrait of an unknown woman, possibly the wife of Francesco del Giocondo. Her pupils look to the right and a corner of her mouth tilts up. We can see her thinking, yet—famously—her thoughts elude us.

Raphael, *Castiglione.*
Courtesy the Louvre
and R.M.N.

Hence an army of observers has declared her everything from angel to lamia. Vasari said her smile was divinely pleasing and deemed her virtual life. In *Revolutions of Italy* (1848–1852), the poet and politician Edgar Quinet praised her "half-ironic smile of the human soul that parades in peace as it looks upon a world liberated from human terror." But Jules Michelet sensed hypnotic evil—"I go to it in spite of myself, like the bird to the serpent"—and in *Italian Art* (1854) Alfred Dumesnil said, "The smile is full of attraction, but it is the treacherous attraction of a sick soul that renders sickness." Walter Pater found the eternal: "She is older than the rocks among which she sits; like the vampire she has been dead many times, and learned the secrets of the grave; and has been a diver in deep seas, and keeps their fallen day about her; and trafficked for strange webs with Eastern merchants." Paul Valéry argued that her face was devoid of mystery. Bernard Berenson came to see a hateful haughtiness in her, and

Leonardo da Vinci,
Mona Lisa. Courtesy
the Louvre and R.M.N.

wanted the painting burned. Paul Ekman of U.C. San Francisco, the world's leading face scientist, believes she is flirting.

One well-known trick used in this painting is sfumato, a blurring of the edges. Sfumato brings a trembling sense of motion to the *Mona Lisa*, especially at the corners of her mouth. It helps explain both the sense of mind and her curious Rorschach effect on observers. Her face is a living multitude.

Certain kinds of character are intrinsically elusive. How does the portraitist show the "real self" of an actor while also showing her acting talent, the knack of feigning a self? In Joshua Reynolds's *Sarah Siddons as the Tragic Muse* (1784), the actress sits with left

Joshua Reynolds, *Sarah Siddons as the Tragic Muse*. Courtesy Huntington Library.

forearm upright like a newel post, and head turned right and slightly up, as if she had just heard a call off-stage. According to Siddons's *Reminiscences*, Reynolds told her: "Ascend your undisputed throne, and graciously bestow upon me some grand Idea of The Tragick Muse." But the pose has bred questions. Did she devise it or was she taking direction from Reynolds? (Stories differ.) Is she even acting here, or, as scholar F. E. Sparshott wonders, simply being herself, "a celebrity dreaming of her own apotheosis"? How much of the sitter do we have?

Intelligence can bedevil the artist. We spot it in people — as a quickly responsive mood of the eye, for instance — but in static faces it grows coy. Portraitists seek the subtly inflected look or the regal glance. Art critics often praise the intelligence in portraits of people like Sir Thomas More, but prior knowledge of the sitter can shape these judgments. Intelligence is a matter of swiftness, grasp of implication, response to nuance, all traits that challenge portraiture.

Intellect almost thwarts the brush, since it requires learning, a quality invisible in faces. How do artists show it? They may depict the subject at mental labor, so in Dürer's 1526 portrait Erasmus sits writing at his table, his face a scowl of concentration. But more often they resort to background, just as lawyers in press conferences place themselves before a wall of books they haven't read. Backdrop seeps into character.

The camera excels at catching character, and also at catching tics and grimaces. A painter can put what she knows of a person into the portrait, but a photographer freezes so many images of face that careful winnowing becomes paramount. A painting takes hours, days, weeks. A photographer can race through a roll of film in minutes, and many shots will show expressions that seem scarcely human.

In the earliest days the camera was slow and required carefully held poses. By 1851 the wet collodion process slashed exposure time to a few seconds, and by the 1860s cameras were fast enough to capture expressions. Nadar (Gaspard-Félix Tournachon, 1820–1910) specialized in catching the personalities of the famous. He made such individuals as Manet, Baudelaire, Daumier, Courbet, and Doré seem distinctive and alert, as if they were thinking or about to speak. We sense their social presence.

Ever since, the camera has excelled at freezing people in poignant moments: Jackie Kennedy's blank loss after Dallas, the defeat and jubilation of players in a World Series, the terror of a Viet Cong with a gun to his temple.

And it has yielded subtler results. Alfred Stieglitz (1864–1946) showed Georgia O'Keefe and Dorothy Norman in a thousand moods. Walker Evans, Dorothea Lange, and Ben Shahn caught the many faces of despair among farmers in the Depression. The great German photographer August Sander (1876–1964) juxtaposed similar individuals to highlight their differences. His 1914 photo of three almost identically garbed men is so rich in nuance that it inspired Richard Powers's novel *Three Farmers on Their Way to a Dance* (1985), which fleshed out their personalities. The photo appeared in Sander's *Face of Our Time* (1929), a collection the Nazis tried to destroy in 1934.

In 1956 Gombrich said, "I doubt if we could ever become aware of the exact changes that make a face light up in a smile or cloud over in a pensive mood simply by observing the people around us," since "we really see a brighter face and not a change in muscular contractions." He was right, and here photography performed another service. It brought science to facial signals.

The Orchestra of Expression

In the nineteenth century, French researcher Guillaume Duchenne de Boulogne (1806–1875) collected bloody heads from the guillotine and touched live electrodes to them, trying to discover how feelings reach the face. He had to work swiftly, since the nerves conducted electricity for only a few hours after death, and he finally tired of this annoyance. So he hired an old almshouse denizen who could not feel pain in his face. The current wrenched his visage into weird and striking shapes, which Duchenne captured on film. By stopping time, he mapped the first reliable geography of expression.

Facial muscles form its base, and we have more of them than any animal on earth, some twenty-two on either side. Like most muscles, they anchor in bone. Unlike most, they attach to the skin. They make facial skin mobile, completely unlike the skin on the back or leg, so it shapes itself quickly to pulses from the brain.

These muscles form a complex skein, as the illustration shows, but some stand out.

The heaven and hell of the face are the *zygomatic major* and the *corrugator supercilii*. The *zygomatic major* is the smile-maker. It slopes down across the cheek to the corner of the mouth, which it tugs upward. The *corrugator* knits the brows together, causing vertical furrows in between. Most pleasant expressions involve the *zygomatic major*, most unpleasant ones the *corrugator*.

Other muscles also play big roles. The busiest is the *frontalis*, the curtain-like muscle of the forehead, which lifts the eyebrows in expressions like surprise. Its opposite number, the *procerus*, pulls the eyebrows down. When we retract the mouth horizontally in fear, we employ the *risorius*, which despite its name plays no role in laughter. The *mentalis* is the pout muscle. It hoists the skin of the chin, pushing the lower lip out.

Most of these muscles are longitudinal, like the biceps. They shrink and tug the skin in one direction. But the face also has two little nooses that float free, not attached to bone on either end. The *orbicularis oris* lies in and around the lips and tightens to pucker them. We use it to kiss or say "who." And the *orbicularis oculi* narrows the eyes. It has two parts, an inner and an outer. The inner closes the eye gently, as in blinks. The outer forces the eye shut, as in a squint or the comic eye-squeeze of children.

All these muscles lie enslaved to the nerves. A single one — the facial nerve — commands them. It has three major parts.

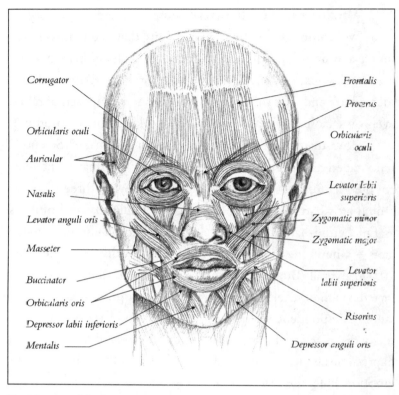

Corrugator

Orbicularis oculi

Auricular

Nasalis

Levator anguli oris

Masseter

Buccinator

Orbicularis oris

Depressor labii inferioris

Mentalis

Frontalis

Procerus

Orbicularis
oculi

Levator labii
superioris

Zygomatic minor

Zygomatic major

Levator
labii superioris

Risorius

Depressor anguli oris

The Muscles of the Face.

The first spurs tears and salivation, and the second bears taste sensations to the brain. The third is the key to expression. Its orders issue from the facial nucleus, a small spot in the pons, an area of the brain above the medulla. It emerges onto the face in front of the ear, where it forks repeatedly to cover the face. It first splits into an upper and lower branch. The upper bifurcates again into the zygomatic and temporal branches, which control expression in the middle and upper face. The lower branch forks into three parts—the buccal, mandibular, and cervical—which move the lower face. Smaller branches exist, but these five are the main marionette strings of the face.

Working together, the facial muscles create a diversity so great we commonly witness expressions that seem novel, yet touch a waiting chord inside. Indeed, it is difficult to make a meaningless expression. The psychologist Paul Ekman sought such a look and after much toil he found one, composed of raised eyebrows, closed eyes, and puffed-out cheeks. It conveys nothing.

How many different expressions can we make? Scientists who taped five hours of interviews with psychiatric patients counted almost 6,000 — that is, a new one every three seconds. Dutch artist Arthur Elsenaar electrically stimulated facial muscles as Duchenne had, and found 4,096 in one 32-minute video-tape — a new one every half-second. Still another estimate cited 10,000 overall. But actor David Garrick could quickly cycle his face through nine feelings: joy, tranquillity, surprise, astonishment, sadness, despondency, fear, horror, despair, and back to joy. How many distinct faces lay in this continuum? Given the gradation of expressions, their mixtures and inflections, and the richness of our emotional life, the real total may defy measure.

Face Esperanto

In 1722 Dutch explorer Jacob Roggeveen was sailing west across the South Pacific Ocean, in quest of a phantom conti-nent. He drifted for weeks across empty water until, on Easter Sunday, he came across a tiny volcanic isle he named after the date. Easter Island was one of the most remote places on earth, a civilization isolated for hundreds of years. But Roggeveen eas-ily read the inhabitants' facial expressions: gladness at seeing him, astonishment at the size of his boat, eagerness to trade.

Every other explorer has had the same experience, in every corner of the world. Columbus, Vespucci, and Cortez had no trouble understanding native countenances, and when outsiders discovered the isolated Biami tribe of New Guinea in the 1930s, they too read their faces instantly.

Face signals are universal. They are a language that reticulates the globe, beyond race, culture, and nationality. Yet for much of this century social scientists believed just the opposite. They thought culture defined facial expressions, as it does words, so they could vary from place to place. Indeed, an anthropological tradition dating from Franz Boas in the early 1900s held that culture determined all behavior. Genes might build the neurons, but society wired them together. We were infinitely malleable. This axiom led Margaret Mead to wildly misconstrue field data in *Coming of Age in Samoa* (1928), and it skewed the work of many other scientists.

It also shaped the assumptions of Paul Ekman at the start of his research. Ekman is a relaxed man with a gray beard, twinkling eyes, and a canny smile as broad as the Cheshire Cat's. At parties people approach him and nervously ask if he can tell what they are thinking. "No, I can't read thoughts," he says, and after a pause he adds, "but I *can* read your emotions."

Partygoers worry about him because he, more than any other modern scientist, has unveiled the secrets of facial expression and deceit. As a child, he loved making funny faces, and his mother sometimes warned, "Don't do that. They'll freeze on you!" In adolescence he became a skilled portrait photographer, and later obtained a doctorate in psychology. When he began investigating face signals in the 1960s, cultural determinism was everywhere. "I expected to show that Darwin was wrong when he said expressions were universal," Ekman says.

To his surprise, he vindicated Darwin. He found that the core of the facial code lies in six glyphs, perhaps a few more. The basic six are: enjoyment, anger, fear, surprise, disgust, and sadness.

These signals emerge in infancy and on a reliable timetable. The smile and surprise appear at birth, disgust and distress (sadness) between 0 and 3 months, the "social" smile at 1.5 to 3 months, anger at 3 to 7 months, and fear at 5 to 9 months. They are almost certainly unlearned. Thalidomide babies born blind, deaf, and armless show them.

For most of us, the basic expressions lie just outside the will. They occur automatically and generally elude facsimile. Most people cannot feign convincing surprise, or even a full smile. Ask them to make a "sad face" and you get a clownlike mope quite different from the look of real sorrow. Since few can exploit it, the face code is believable.

The most easily recognized expression on earth is the true smile, the flare of happiness. It employs two muscles. The *zygomatic major* curves the mouth and the *orbicularis oculi* hoists the cheek, pressing skin toward the eye in a squint of joy. In a wider smile, the teeth flash and the eyes glisten. Not every smile indicates pleasure, but the true smile is unmistakable.

We describe the smile easily, but the other portraits in this gallery are more elusive. What, for instance, is the face of wrath?

Anger is a look of dark concentration. The face seems to contract to aim its malice more intensely. The eyebrows descend, and the lips tighten. Jane Eyre says of her Aunt Reed's eyebrow, "How often had it lowered on me menace and hate!" Blood rushes into an irate face, flushing it. In rage, the eyes gleam and the countenance can seem to shine. An angry face is a warning. Like the brilliant colors of a coral snake, it helps minimize physical conflict, which can harm even the victor. It gives the target a chance to appease.

Fear has almost the opposite characteristics. It seems to open the face. The eyes widen, and the eyebrows lift and move toward each other. The lips pull back horizontally. As fear worsens, the

The True Smile. Photo: Paul Ekman.

lips may tremble. The mouth dries and can open and shut, the source of Jackie Gleason's famous *Honeymooners* fear-babble: "Hummunna hummunna hummunna." In terror, the face blanches, the nostrils dilate, the pupils widen enormously, and perspiration dots the forehead, the legendary "cold sweat." On the other hand, in slight fear, one may yawn. Fear is the complement to anger. It suggests appeasement and thus reduces pummelings.

Surprise resembles fear and often precedes it. It is the briefest expression, lasting less than a second. Both the eyes and mouth fly open, and the eyebrows rise and arch. Hence in Italian *fare tanto d'occhi*, "to make the eyes so big," means "to be amazed," and Ray Milland called his autobiography *Wide-Eyed in Babylon*. The degree of brow-lift and jawfall registers the amount of surprise. But they must work together, for a wide-open mouth with barely raised eyebrows is meaningless, a gape.

Anger. Photo: Paul Ekman.

Surprise seems to have begun as self-protection. When animals like dogs are surprised, they flatten their ears to keep them safe. In humans, the muscles for ear flattening have migrated to the forehead, where they hoist our eyebrows. And we open the mouth in surprise, Darwin felt, because we breathe more silently that way, a boon in danger. Dogs, he noted, breathe more quietly through their noses, so they shut their jaws when alarmed. An open mouth also prepares us to draw large drafts of air, vital in an emergency. Astounded chimps and orangutans also open their mouths, and often protrude their lips as well.

Disgust centers on the nose, not otherwise a marvel of expression. It may partly turn up, wrinkle, and contract, as when we sense a revolting odor. The face is using a metaphor: the loathed item stinks. "Disgust" stems from Latin roots meaning "not tasty," and the mouth comes into play as well. We may

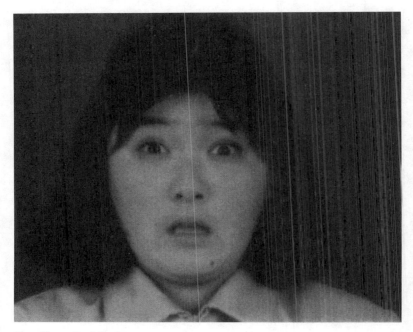

Fear. Photo: Paul Ekman.

show moderate disgust by spitting and by opening the mouth as if to eject a chunk of putrid food. We convey extreme disgust with mouth and throat movements like those before vomiting.

These reactions make sense, since disgust is the emotion that guards the mouth. The most disgusting sensations involve excretions and decay, and almost all stem from animals or animal products. Some people dislike eating tongue because it is so plainly animal, and sashimi revolts others for similar reasons. In fact, the link between odious tastes and the disgusted face is anatomical: The same part of the brain reacts to both.

Disgust flows like electricity through associations. Diners won't eat soup stirred with a brand-new fly swatter or comb. Upper-caste members in India balk at food handled by an untouchable. In one hospital, thirsty nurses were drinking glasses of juice meant for children, so administrators began serving the

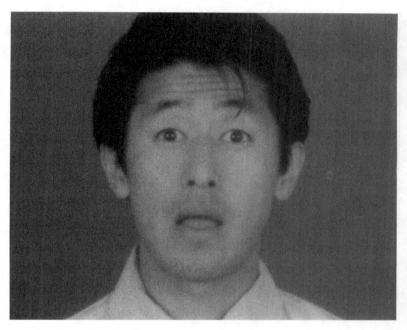

Surprise. Photo: Paul Ekman.

juice in brand-new urine containers. The problem stopped. This extraordinary conductivity probably arose because even invisible filth can harm us. Scientists have speculated that the disgust expression arose to teach cultural values, by associating taboos with revolting objects. Indeed, once we learn disgust of certain foods, like shellfish, we rarely unlearn it. The disgust face also warns others of sickening foods or odors.

Sadness makes the face seem to sag. The eyebrows drop, though their inner corners rise slightly in a shallow pediment. Wrinkles pool in the mid-forehead, and hence many writers have noted that grief dwells in the brow. The mouth droops into a frown and the features go slack. We develop a "long face." When Indians on Tierra del Fuego wanted to tell Darwin that a sea captain was depressed, they pulled down their cheeks. This expression resembles those for distress and guilt.

Disgust. Photo: Paul Ekman.

A sad face evokes compassion and aid, even among chimps. Indeed, it is so effective that it can become a stylized plea. Professional beggars specialize in long faces. Among the Kaluli of New Guinea and the Quechua of highland Ecuador, a look of sorrow is not just a momentary signal, but a well-defined cultural message. When a Quechua woman grows sad for a long period, her kin investigate, usually suspecting the husband of misdeeds.

To these six signals, some scientists add contempt, pain, and neutrality.

Contempt is a subtler display, and scientists have argued about not only its universality, but its lineaments. The experiments that have shown cross-cultural recognition use a classic lip curl or sneer, with the lips tightened and lifted on one side. But others have invoked a lofty attitude of eyes, and Darwin suggested the eyelids partly close and turn away, as if unwilling to

Sadness. Photo: Paul Ekman.

behold the scorned object. Possibly contempt has several expressions. The dispute is lively and, on paper at least, reasonably free of contempt.

Pain may be a basic expression. We're all familiar with the face "contorted" in pain. The look is riveting and can be almost unbearable. The central face seems to squinch. The eyebrows narrow and often drop, and the bridge of the nose wrinkles. The mouth opens as the upper lip rises, and wrinkles may appear beside the nose. In agony, the eyelids may shut completely. The face closes in on itself, as if fending off invisible harm. We don't even begin to show this display unless the feeling is intense. Hence observers underrate pain based solely on expression. People who have lived with a suffering individual assess it better, while clinicians surrounded by pain discount it even more, an important fact they may not realize.

Contempt. Photo: Paul Ekman.

Pain wins aid even more dramatically than sadness. It also highlights the social and evolutionary aspect of facial signals. We feel small amounts of pain without any change in expression. The signal appears when pain is harsh, just when we might need help from others.

Some scientists have suggested neutrality is a facial expression, the default. We all recognize a neutral face, but it is a strange expression, since we can't read the mind-state behind it. In fact it is opaque, the absence of expression. One can be happy, for instance, and have a blank face. In one study experimenters showed American and Japanese subjects gruesome movies — of ritual circumcision, suction-aided birth, and nasal sinus surgery — and some 20 percent of them displayed a neutral face throughout. What were they feeling? We can only guess.

To some extent, the basic expressions are a periodic table of the face. They can alloy, so we see scornful anger, enjoyable anger, disgusted anger. They also have isotopes. Anxiety, which is worry about impending harm, resembles moderate fear, mainly in the horizontal mouth stretch. Attention raises the eyebrows slightly, like mild surprise. Distress and guilt seem spinoffs of sadness.

But the analogy to the periodic table is misleading and quickly breaks down. For one thing, facial displays change along a gradient. Anger varies smoothly in intensity (from irritation to fury), level of control (from tense fuming to red-faced explosion), and genuineness (from feigned to real). More extreme expressions, like terror, often transcend the prototype and boast special idiosyncrasies. And the sheer number of variants in some cases, like disgust, make the basic expression seem a swarm.

And the gallery may not be complete. It omits joyous laughter, yawning, and blushing, all universal acts and highly expressive. It does not address an array of other mind-states, like bewilderment, boredom, sleepiness, and suspicion. They seem universal—we recognize them in subtitled movies—but they have evaded experimental scrutiny.

As scientists admit, the research here suffers some of the same pitfalls as in physiognomy. It typically relies on still photos, which tend to use a head-on pose that flattens the face, robbing it of depth cues. They often show constructed expressions rather than everyday ones.

And they miss flux. Expressions are dynamic, with onsets, peaks, and die-aways lasting several seconds. Yet photos are cross-sections in time. They catch perhaps 1/500th of the whole expression. Taken at the peak moment, they may capture a signature look recognized worldwide, but too early or late and they can fail. And some expressions, like the stare, may depend so much on time they elude still photos entirely. Hence some researchers have begun working with videotape.

In *Brazil*, John Updike writes, "When Isabel delivered herself of a remark meant to prompt a response, her whole face showed tension, a kind of bright brimming, as of a plump dewdrop about to break and run." The face abounds with such vivid signals. What constitutes them? We don't know.

The Wink and the Baton

The basic expressions are the lords of facial signals. They reveal deep and often passionate feelings, but except for the smile they are fairly rare. Beneath them and flitting about incessantly are messages of a different sort: paralanguage. Paralanguage includes word substitutes, inflections, talk coordinators, and even semi-noise, like working the jaw. These are the buzz and bloom of expression, the little signals that constantly tie us together.

Though we use them incessantly, scholars did not start examining them until the 1950s. Paul Ekman and Wallace Friesen later classified paralanguage into four kinds: emblems, illustrators, regulators, and adaptors. They provide an *omnia Gallic* to the territory.

Emblems replace or modify spoken words. They stand in for utterances. A smile can state: "I agree." A hoist of the eyebrows can say: "I doubt it." A shrug, along with a brief downturn of the mouth corners and a slight tilt of the head, means: "I'm baffled." A wink can refer to shared experience, as in merry collusion, or convey flirtation, or indicate a comment is not meant seriously. A stuck-out tongue is an insult. In the United States we currently use about sixty different emblems.

These are conscious signals. People are usually aware of using them and not aware of their emotional expressions, which are not paralanguage. Hence we often respond to an emblem as we would to a sentence. For instance, if we make a claim to someone and she gives a skeptical brow-lift, we may answer it just as if she had stated, "I don't believe you."

Emblems are a fine conduit for the sub rosa. Ask a person his opinion of someone in the next room and he may grimace. Like a whisper, this tactic narrows the message to the visual. No one can overhear it. It's also deniable, since he hasn't uttered repeatable words.

These signals often ape the inborn ones. But emblems tell; the basic expressions show. We don't feel the emotion (except as memory recalls it), but refer to it. If someone asks your opinion of a political candidate and you wince, you aren't saying, "I'm in pain," but "He gives me pain." Such messages are often shorter or longer than the expression, and usually stylized, either more or less mobile than the original.

Emblems are learned and vary among cultures. In the Chinese novel *The Dream of the Red Chamber*, scratching the ears and cheeks indicates happiness. Polish nobles of the sixteenth to eighteenth centuries developed emblems to distinguish themselves from rabble. Two equals might greet by kissing each other on the shoulder. A man receiving appreciation from a superior might kneel and have his head hugged. Anyone ignorant of this extravagant code stood forth as an imposter.

Emblems need not involve the face. The upthrust middle finger is an emblem widely recognized in Western cultures, and the ancient Romans understood it too. Virtually all hand gestures are conscious and culture-coded, perhaps because, unlike the face, we can see the arms completely and thus control them better.

We sometimes use emblems to overcome distance, noise, or protocol. A waving hand can convey "hello" between two faraway people. Sawmill workers, who labor in a metallic din, have intricate hand-gesture codes, and silent monastic orders like the Cistercians have used sign language at least since 328 A.D. In the deferential silence that cocooned the Ottoman sultans, palace pages developed a similar code. Of course the deaf speak with full-scale manual languages. Indeed, in American Sign Language facial expressions ascend from paralanguage to real language. For instance, an expression with lips pushed together and slightly out, and head slightly tilted, can mean "with relaxed enjoyment."

The illustrator is the second area of this province. Whereas the emblem is a whole comment, the illustrator more resembles an underscoring or fingerpost. The subtype called the baton emphasizes a word or phrase. For instance, raised eyebrows with "What do you *want?*" is a baton. We see such italics every day, and they are one more benign fifth column of the face, sneaking past our conscious awareness to help us within. Indeed, the brow-lift is the most common facial display.

A few people use the normally involuntary sadness/fear brow movement—lifting the inner eyebrow—for this purpose. Woody Allen is one, and Ekman says this gesture contributes to his wistful mien. The realtor Gerri in Stephen Wright's *Going Native* (1994) is another, and the trick gives her "a perpetually bemused look she employs to her benefit, coaxing empathy and contract signatures from wavering clients.' Raising the eyebrows also can signify exclamation or query.

Some illustrators show location. The face can be a pointer, as when we move our eyes and tilt the head: *That's him over there.* We use illustrators for more general spatial relationships too, lifting or lowering the brow, for instance, to show up or down.

The regulator is the third area. Regulators manage the flow of conversation. They are like the traffic signal at an intersection, but come in wide variety and can be virtually subliminal. These subtle signs prevent collisions and abyssal gaps in talk. To show desire to speak, for instance, we move the mouth in silent, hesitant mimicry of interruption. We raise the eyebrows to concur and tell the speaker to continue. To convey the opposite — we disagree and want her to stop — we may frown and shake our heads. To prod her to hasten the narrative, we nod staccato-style. And mutual smiles lubricate the entire process. On the phone, without regulators, we interrupt much more.

The final kind of paralanguage is the adaptor. Adaptors are self-manipulations and lie out at the gray edge of signaldom. They include random lip-biting, lip-wiping, running the tongue between teeth and cheeks, jaw-working, eye-clamping, and a variety of weirder contortions, like sucking the lips into a figure 8. They are the burps and gurgles of facial expression.

Paralanguage is a microcosmos, a little realm of spinoffs from the grand and profound signals of the face. Yet they spice and streamline our conversation, and without them the face would seem much emptier.

The Speaking Eye

In Balzac's short story "The Elixir of Long Life," a dying man asks his son to spread a magic potion over his corpse to resurrect him. The youth applies it first to one eye. It opens, and all at once it is "thinking, accusing, condemning, threatening, judging, speaking, shouting, and biting." The son shuts the eye and keeps the elixir for himself.

The eyes are supremely expressive. Pliny the Elder said, "No other part of the body supplies more evidence of the state of mind." Jane Eyre muses, "The soul, fortunately, has an interpreter — often an unconscious but still a faithful interpreter — in the eye." Gerty MacDowell's soul is in her eyes. Explorer/

scientist Mary Kingsley (1862–1900) noted that tribes in Central Africa often rifled graves for eyeballs, seeking to "secure 'the man that lives in your eyes' for service to the village."

Many expressions too subtle for the basic schema lie in the eyes. Indeed, the eyes convey thought. Film actors know their art lies mainly in the eyes, which show the course of their minds. The eyes are the center of expression for cartoonists as well, who control them by adjusting their position on the face, especially in relation to the eyebrows.

The eyes seem capable of infinite range and finesse. They can hold one's history. "Looking into his eyes," says Ishmael of Starbuck in *Moby-Dick*, "you seemed to see there the yet lingering images of those thousand-fold perils he had calmly confronted through life."

The eyes can mock, scorn, and challenge. In H. Leivick's *The Golem* (1921), the half-formed clay man fears people's glances. Certain murderers report having killed because they didn't like the way the victim was looking at them, and among British soccer gangs, a glance from a rival group member can spark a brawl, often heralded by the cry: "He looked at me!"

The eyes can convey amusement. The Japanese call it *mokushoh*, or "eye-laughter." Frans Hals's *The Laughing Cavalier* (1624) isn't laughing, or even technically smiling. But his mustache curls up like a smile, his eyes seem to twinkle, and his whole face conveys merriment. He is laughing with his eyes.

Eyes are erotically eloquent. When Lady Booby first tries to seduce Joseph Andrews, she asks if he's ever been in love. He falters and she purrs, "Tell me truly, who is the happy girl whose eyes have made a conquest of you?" In boarding school, the gaze of Becky Sharp besots a clergyman and the headmistress can scarcely believe she hasn't spoken to him. Malinowski said the Trobriand Islanders, who went about naked, called the eyes "the gateways of erotic desire," and spent more time decorating them than any other part of the body. High-level Tantra practitioners locate attraction not in sexual tug, but in the "meeting of eyes."

Love at first sight usually involves eye contact, and the Greek poet Meleager (1st cen. B.C.) placed Cupid inside the eyes, whence he aimed his barbs. In much subsequent love poetry, and later among Arab poets and Provençal versifiers of the twelfth century, the eyes shoot arrows, darts, or fiery beams that wound the soul and infect with longing. In *La Vita Nuova* (c. 1292), Beatrice flashes love from her eyes and inflames Dante. Neo-Platonist Marsilio Ficino (1433–1499) believed that blood mixed with emanations from the eyes and thence flew in a glance to the other person. Robert Burton, in *The Anatomy of Melancholy* (1621), adopts Ficino's theories and says people "direct sight to sight, join eye to eye, and so drink in and suck love between them."

Ficino's charming idea has capsized in the seas of science, but romantic eye-beams remain as powerful as always. In Zola's *Une Page d'Amour* (1878) Hélène and Delberte fall into deep eyelocks that pale words: "Both told each other with a look that they would love each other here, or there, wherever they might be together." A world away, on the rainy island of Chiloé in southern Chile, lives an elfin character called the *trauco*. He is short, but women out wandering the fields must be alert, for he can seduce them with his magic eyes.

In strict Muslim countries where women shroud the whole body except the eyes, the eyes become the face. Guy de Maupassant said Europeans "often become attached to these indolent and enveloped creatures, whose eyes alone can be seen." Darwin saw women in Peru whose veils left one eye exposed, and felt it was magnificently expressive. Recently in Saharan Algeria, traveler William Langewiesche saw between the folds a female eye of such inviting beauty that, he says, "I didn't need to see more."

A sparkle in the eye is especially fetching, and portrait photographers try to show a single gleam in each, the "catchlight." Henry Fonda always had a tiny light called an "inky-dink" placed near him in close-ups. He gazed right into it and it made his orbs glow.

Eye expression is slippery terrain, and few scientists have so far scouted it. One is Johnmarshall Reeve of the University of Wisconsin, who found the eyes crucial to the expression of interest. When intrigued, people widen the eyelids, exposing more of the eyeball, and they may part the lips slightly. At the same time, they make few eye glances or head turns, and generally keep the head still. In his experiment, people identified the face of interest from film clips, which are probably essential to this kind of research.

How many kinds of messages can the eye convey? We don't know. Psychologist Simon Baron-Cohen suggests fifteen spectra, in addition to interest: trouble and calm, tenderness and callousness, certainty and uncertainty, reflection and thoughtlessness, seriousness and play, sadness and happiness, attention and inattention, dominance and submission, friendliness and hostility, desire and hate, trust and distrust, alertness and fatigue, falseness and sincerity, surprise and knowledge, and anger and forgiveness.

Indeed, eyes are so evocative that in literature they become home for mind-states that otherwise thwart description. Iago blandly warns Othello that jealousy is "a green-eyed monster which doth mock the meat it feeds on." In China, covetousness is "red-eye disease."

Poet George Herbert (1593–1633) said, "The eyes have one language everywhere," and its vocabulary may be just emerging. Baron-Cohen notes we sense a variety of factors. Movement alone can communicate. Eyes lifted upward and at an angle can convey skepticism. Speed of eye motion is also revealing. A slow glance away can indicate boredom, a quick one desire to conceal. Other likely cues are the size of the pupils, their position relative to the whites, the position of the eyelids and brows, and the focus of the eyes. Baron-Cohen tentatively suggests that the train of eye movements can resemble words in a sentence, complete with syntax and richly nuanced meaning. "In a real sense," he says, "there may be a language of the eyes."

Ask most people why we have eyebrows and you'll get a blank shrug. But as Cicero realized and joggers know well, they keep perspiration from falling into the eyes. It is a subtle but serious function, given the dense stippling of sweat glands on the forehead, and it likely proved very useful to hunters on the savanna. It even helps nonathletes on hot summer days.

Eyebrows also act as ornamental thatches, drawing attention to the eye. Like the lids and lashes, they affect it as aides do a head of state, making it look better while remaining discreetly on the sidelines.

Intriguingly, not all cultures have viewed eyebrows as an esthetic prize. Fashionable women in Chaucer's England plucked them, and Hanafi Muslim women do the same in the *hamam*, or Turkish bath, partly for religious reasons. Many Amazon tribes — the Bororo, Canella, Mundé — deem eyebrows hideous and pull both them and eyelashes out. The Mbaya of the Amazon remove all facial hair, including eyelashes and eyebrows, and sneer at full-browed whites as "ostrich-brothers."

But sweat deflection and orb adornment are ancillary. By now we've seen the most common use of eyebrows again and again: They communicate.

The eyebrow is the great supporting player of the face, and its work generally escapes notice. It helps signal anger, surprise, amusement, fear, helplessness, attention, and many other messages we grasp at once. Indeed, without eyebrows the surprise expression almost disappears. The eyebrows are such active little flagmen of mind-state it's amazing anyone can wonder about their purpose. We use them incessantly.

For instance, when we greet a friend at a distance, we quickly raise and lower the eyebrow: the brow-flash response. It lasts about a sixth of a second, and conveys surprise and delight. Researcher Irenaus Eibl-Eibestadt found it among Europeans, Samoans, !Kung Bushmen of southern Africa, and the Quechua of Peru, and it seems universal — yet almost no one is consciously aware of it.

The eyebrows go much beyond the basic expressions. For instance, when we pull them down and together, the result suggests perplexity, concentration, attempt to overcome a problem. Hence Darwin called the *corrugator*, which performs this act, "the muscle of difficulty."

"The eyebrows form but a small part of the face, and yet they can darken the whole of life by the scorn they express," said the orator Demetrius (c. 350–c. 283 B.C.). Though distinct from classic contempt, a lift of the eyebrows clearly conveys skepticism, most likely by its resemblance to surprise. A speaker strays into absurdity and receives a facial "Oh, really?"

French poet Maurice Scève (1510–1564) wrote that the eyebrows have a godlike power over lovers, and certainly they help flirtation. Says Eibl-Eibestadt, "The flirting girl at first smiles at her partner and lifts her eyebrows with a quick, jerky movement upward so that the eye is briefly enlarged." She looks abruptly away, then slowly back. The gesture is identical in men.

Admiration lies in the brow. Darwin called it surprise plus pleasure and approval. In the more intense cases, he said, the eyebrows rise, the mouth smiles, and "the eyes become bright, instead of remaining blank, as under simple astonishment."

The eyebrow can convey helplessness. When we shrug our shoulders, we raise our eyebrows as well, and we're usually unaware of it. But a shrug seems meaningless without a brow-lift. Shoulder-shrugging may be innate. It appears worldwide and Darwin cites a blind woman who did it.

The great eyebrows have all excelled at expression. Groucho Marx's huge false eyebrows comically emphasized his quicksilver whims and lusts. John Belushi was renowned for his arch of a single eyebrow, creating an intense, quizzical stare. Sam Ervin's orbital tufts fluttered famously in indignation as he questioned Watergate conspirators. On the other hand, Uriah Heep had hardly any eyebrows and no eyelashes. It's fitting. His face is all falsity.

Facial Zemzem

The source of tears long mystified humanity. The ancient Egyptians thought them a brimming-over of the heart. Hippocrates said they descended from the cerebrum, a literal brain drain, and Plato believed them a mix of the eye's moisture with "visual fire." Galen (c. 130–c. 201) held that tears arose from nearby glands, but the medieval Arab scholar Hunain (9th cen.) thought they came from the brain when people sneezed. The versatile Niels Stensen (1638–1686) solved the problem in 1662. By dissecting corpses, he discovered the main lacrimal gland, just above the outside corner of the eye.

In most animals tears simply keep the eyeball wet, but in people they have evolved into one of the most dramatic of all expressions: weeping.

We link tears to grief, and like a melancholy song, tears can have eerie beauty. As the sisters of Phaeton wept for him, their tears turned into amber. In Pynchon's *The Crying of Lot 49* (1966), Oedipa imagines her wraparound shades filling with tears, so for the rest of her life she sees the world through that moment's pain. Mandeville describes a lake high in Sri Lanka where Adam and Eve wept for a hundred years after their ouster from Paradise. Their tears formed the lake and gems bestrew its bottom.

Tears tend to arise from an emotional fault-drop, or "sudden dejection," as Thomas Hobbes (1588–1679) called it. They stem, he noted, from acts that abruptly remove "some vehement hope, or some prop of [one's] power; and they are most subject to it, that rely principally on helps external, such as are women and children." Even a suddenly enhanced *sense* of loss can spur tears. At funerals, when a minister praises winning traits in the deceased, mourners can begin sobbing.

Culture strongly conditions tears. Darwin observed that the English wept less than Mediterraneans. In the nineteenth century people cried reading Dickens and patrons of Bayreuth learned to sniffle appropriately. In Poland, noble males wept freely, and when King John Casimir heard of a military triumph, tears the size of peas apparently coursed down his cheeks. But today in Western cultures tears are unmanly and we find Stan Laurel's naive weeping hilarious. Women cry longer and more often than men. In one study they wept 5.3 times a month, compared to 1.4 for men.

Babies cry irritant tears, but emotional ones appear at several weeks to three months. An average one-year-old cries 65 times per month. Some scientists suggest these cries began as a tool of infant tyranny. On the savanna they attracted predators, so they bent parents quickly to the toddler's will. Even today they retain a note of urgency.

Ella Wheeler Wilcox (1850–1919) wrote, "Laugh, and the world laughs with you;/ Weep, and you weep alone." This line, now a maxim, is literally quite false. Depression isolates people, but weeping brings them together. A cry-face in a child evokes concern in parents, and in an adult, compassion from others. The pull is arresting. A person sobbing in a supermarket, say, generates an instant circle of wellwishers. Handling a baby chick, Lady Chatterley suddenly bursts out crying "in all the anguish of her generation's forlornness," and the keeper touches her shoulder, a gesture of comfort that begins their affair. Ovid tells wooing men to shed tears. "They move the most adamant natures. Let her, if possible, see tears on your cheeks, in your eyes." And if you can't produce tears at will, he says, bring a vial of water and fake it.

Grief is not the sole cause of tears. In *The Epic of Gilgamesh*, the storm god brings darkness and flood to the land, and the gods cower, weeping. Yet Ut-napishtim, the Mesopotamian Noah, also weeps when the downpour ends. We cry at tragedy as well as its relief. Hard laughter can yield tears. So can gratitude, and in *The Seagull* (1896), the tense young Treplev's eyes fill with tears when the local doctor praises his play. At night in boarding school, Becky Sharp weeps in rage. People may also cry from anxiety, fear, pain, and weariness.

Glory, success after extreme effort, can bring tears. Empress Eugénie cried when she saw the first ships sail through the Suez Canal. It is one secret of a Hollywood weepie. At the end of films like *Rocky*, the underdog usually not only attains the goal he has dreamt of and suffered for for two hours, but more. The glory rushes in almost too quickly to handle, and viewers' tears well up.

Hobbes noted that, for some, weeping can stem from "the sudden stop made to their thoughts of revenge, by reconcilia- tion." Vaporizing fear of reprisal can also spur tears in the mul- tiplex. Movies ending in interracial grins and hugs often yield tears, at least for whites.

Perhaps oddest of all, music can cause tears. One duct- squeezing trick entails repeating a musical theme a step higher or lower than when the listener first heard it, as in Albinoni's *Ada- gio for Strings*. Another, even more effective, is the appoggiatura, or delay in resolving a musical theme. The Beatles' "Yesterday" begins with one, and appoggiaturas abound in tearjerking tunes. They build a kind of anxiety, which the melody finally resolves.

We are the only primate that cries. In apes tears simply keep the eyeball moist. However, as Elaine Morgan has pointed out, sev- eral aquatic species shed tears. Sea birds release them through their noses, and these tears are high in salt. Sea crocodiles shed tears; river crocodiles don't. Seal mothers shed tears when sepa- rated from their young, and some birds may show a similar reac- tion. Weeping is another prong in Morgan's argument for an aquatic phase of human evolution.

Human tears come in two kinds, emotional and "irritant," which differ in chemical composition. Biochemist William Frey found that emotional tears have 21 percent more protein and suggested that they carry away chemicals the body produces in stress. That's why a "good cry" makes us feel better.

This notion fits the evidence. Tears are triggered by the parasympathetic nervous system, which regulates maintenance, rather than the sympathetic, which handles emergencies. More-over, any high emotion—joy or sadness—could require a sub-sequent cleanup, a quick purge of lingering chemicals.

Hence the tears of Empress Eugénie. The stress of under-taking the Suez Canal had finally ended, and her body rid itself of suddenly unneeded chemicals. Likewise at the end of a *Rocky*, when tension vanishes in a golden haze of glory, the body sud-denly does not need stress chemicals, and we force them out in tears. We literally shed our anxiety.

Carnival of Smiles

In *War and Peace*, Natasha's smiles almost speak. When Prince Andrei asks her to dance, her smile seems to say, "I've been waiting an eternity for you." At the ball where she charms the prince, her father asks if she's enjoying herself and her reproach-ful smile says, "How can you ask such a question?" When another dancer sweeps her away from Andrei's side, she smiles to him. "'I would rather rest and stay with you, I'm tired,' said that smile; 'but you see how they keep asking me, and I'm glad of it, and happy, and I love everybody, and you and I under-stand all about it.' That and much more this smile of hers seemed to say."

As anatomist Charles Bell pointed out in 1806, a smile can convey a thousand different meanings. Yet it is also the most easily recognized expression. When a tachistoscope limits expo-sure to a fraction of a second, people spot smiles far more than other expressions. Only smiles and surprise are identifiable in faces at 150 feet, and only smiles at 300.

Smiling is innate and appears in infants almost from birth. Babies born blind and deaf will smile, though they've never seen smiles themselves. The first smiles appear two to twelve hours after birth and seem void of content. Infants simply issue them, and they help parents bond. We respond; they don't know what they're doing.

The second phase of smiling begins sometime between the fifth week and the fourth month. It is the "social smile," in which the infant smiles while fixing its gaze on a person's face. Voice and touch can also elicit it. The "social smile," however, still seems ill-defined, an obstacle to fixing its onset.

Human smiles resemble the primate "grimace" or "silent bared-teeth face," in which apes retract the corners of their mouths and show their teeth. It sometimes appears when two animals meet after a time, but only in the subordinate. It also arises when an ape is under attack or threat, and in males during copulation.

Is this the origin of the smile? Does it begin in fear and prostration? A few scientists think so, but others note that the human smile uses the *zygomatic major* and the primate grimace doesn't. They suggest, rather, that the human smile stems from the primate "play-face" or "relaxed open-mouth face." This expression, in which the mouth opens wide and the corners pull back only slightly, resembles the human smile less. But it occurs in similar contexts, uses the same muscles, and at times can suggest a person's grin.

People seem to return a true smile automatically, even when just looking at photos. We like smiles, which is why we see so many from TV ads, salespeople, and often strangers. Dale Carnegie said smiles can win friends and influence people, and indeed studies show we deem smiling people more pleasant, sociable, attractive, able, and honest than unsmiling ones. Though courtroom judges are equally likely to find smilers and nonsmilers guilty, they give smilers lighter penalties, a phenomenon called the "smile-leniency effect."

People often indicate compliance, intent to cooperate, with smiles. The smile may be the expression most easily recognized across cultures because it best fosters cooperation. It may be the most feigned expression for the same reason.

Indeed, much of expression goes to affirming good will. Without it suspicion can brew. For instance, the world of online chat lacks the backup we get from face and voice, so sometimes participants find sarcasm in the genial line, rivalries fester, and arguments erupt—all over vapor. Hence the sideways faces built of punctuation marks: emoticons. They convey tone, and a smile, such as

$$8 \star) \qquad \% \wedge) \qquad \# : -) >$$

is by far the most common. It says: I mean well.

Yet smiles are also one of the most ambiguous expressions. Indeed, they vary like a kaleidoscope. Turn the tube slightly, change a nuance here or there, and a new meaning arises. For instance, among the smiles in Japanese are: *niko-niko*, a smile of peacefulness and content; *nita-nita*, a smile tinged with contempt; *ni*, a brief grin; *niya-niya*, an often unpleasant way of smiling when suppressing joy; *ninmari*, a smile after achieving a goal; *chohshoh*, a sneer. "The very word 'smile' is problematic because it causes us to group into one category things that are different," Paul Ekman says. "Your smile when you go in for root canal work is entirely different from your smile when you get a raise."

Here are some smiles Ekman has charted:

The *felt* or *enjoyment smile* is the standard. Thomas Hardy describes it in *Far from the Madding Crowd* (1874): "When Farmer Oak smiled, the corners of his mouth spread till they were within an unimportant distance of his ears, his eyes were reduced to chinks, and diverging wrinkles appeared round

them, extending upon his countenance like the rays in a rudi-
mentary sketch of the rising sun."

But contempt also involves a smile. As we've seen, the lips
tighten and one corner tends to ride up. This is the smile of
Snidely Whiplash and many movie villains. Since contempt can
itself be pleasurable, it can merge with the enjoyment smile and
blur the distinction between these signals. The *Mona Lisa* has a
raised mouth-corner, and it may partly explain the malice some
find in her.

Fear and pain yield smiles, though curious ones. In fear the
risorius muscle pulls the lips back and its corners can tip up,
smile-like. In pain, the corners can also rise, in a cringe. Neither
has anything to do with enjoyment.

Beyond the basic expressions, smiles get more complex.
When a person tries to hide enjoyment, as when delighted by
harm to another, she may show a *dampened smile*: a smirk. The lips
may press tightly together, as if trying to hold the smile within.
The mouth corners may tighten, the lower lip may ride up, the
lip corners may pull down, or any combination may occur. Yet
the eyes usually squint with pleasure, sabotaging the effort.

The *miserable smile* is very different. It is the grin-and-bear-
it display, suggesting that the person feels wretched but will not
complain about it. It is commonly asymmetric and often appears
atop negative expressions, not hiding them, but adding to them.
It differs from the smirk mainly in that the eye muscles don't
tighten. There is no joy.

In the *qualifier smile*, the speaker seeks to cushion delivery
of an unpleasant message, and it may induce the recipient to
smile back. It begins abruptly. The bearer of bad news tightens
her lip corners and may push her lower lip upward for a
moment. Sometimes she nods her head and turns it slightly
down and sideways, looking down upon the other person.

In the *compliance smile*, the listener offers the cushion. This smile says that the person will accept bad news without protest. Oddly, it looks like the qualifier smile, but instead of moving the head, the stricken individual may raise the brows for a moment, and may sigh or shrug.

Some smiles flicker about as paralanguage. The *coordination smile* is a slight, usually asymmetric smile that denotes agreement, understanding, or intent to perform between two people. It does not involve the eye muscles. The *listener response smile* amounts to an "um-hmm," a notice that the smiler comprehends and the speaker can proceed. A head nod often goes along with it.

Smiles merge with gazes to create more intricate signals. For instance, in the *flirtatious smile*, the flirter briefly shows an enjoyment smile, gazes away, then back. In the *embarrassed smile*, a person smiles but may gaze downward or away. Smiles tend to disarm aggression—one reason we smile when meeting new people—and the embarrassed smile may help deter attacks at a vulnerable moment.

And of course smiles, especially the enjoyment smile, can fuse with anger, contempt, fear, sadness, and excitement to create many distinctive blends.

How many different smiles are there? Ekman has described eighteen exemplars, and their varying alloys, degrees, and contexts yield far more. Yet, intriguingly, few grand unified theories of smiling have appeared. With laughter, the case is different.

A History of Theories of Laughter

Lord Chesterfield once said that a gentleman should never laugh. It was enough to smile graciously. He should probably have said a gentleman never theorizes about laughter. The quest to explain why we laugh has gone on since Plato, and it has not been dignified.

The sheer variety of opinion is impressive. Humor shows aggression, says one observer, and masochism, says another. One says laughter comes from misery. Another says it comes from joy. A third says it comes from joy and misery felt together. An untickled child will grow up laughless, we read. Dogs laugh by wagging their tails. Even microbes have a sense of play.

Famous thinkers have struggled here. The puzzle has lured Descartes, Rousseau, Priestley, Hegel, Hazlitt, Emerson, Dewey, and more. A few, like Schopenhauer, claimed to have solved it, and many have dealt with it breezily, in a few paragraphs or even sentences. The careful admit to tiptoeing through this zone. Dewey warned that few would believe his theory, Freud approached the comic with misgiving, and Darwin said, "The subject is extremely complex." Great comedians like W. C. Fields have confessed their utter ignorance of it, and Groucho Marx once said, "I doubt if any comedian can honestly say why he is funny and why his next-door neighbor is not."

There are now over one hundred theories of laughter, and with a little folding and amputation they fit into a few over-lapping types. Indeed, they lean so much on each other that they are less separate theories than facets of a whole. They are: superiority, incongruity, surprise, ambivalence, release, configuration, and economizing.

> *He was born with a roaring voice and it had the trick of inflaming half-wits*
> H. L. MENCKEN,
> on William Jennings Bryan

Superiority theory holds that we read this jest and laugh, if we do, because we vault pleasurably above its target. This notion began with Plato, who noted that malicious people laugh at the misfortunes of others, and Aristotle, who said we find the ridiculous in the painless, harmless defects in others. Hobbes described the source of laughter as "sudden glory." Another's social slippage makes us applaud ourselves in comparison, and the less secure we are, the harder we laugh. Sudden glory has a tinge of ignominy.

Voltaire disagreed: "In laughter there is always a kind of joyousness that is incompatible with contempt or indignation." Yet others' mishaps can in fact bring merry tears to our eyes. Unfortunately for the theory—and fortunately for the species— they don't always. A dull, uninventive insult is rarely funny and often dislikeable. And research shows that the worse the plight of the butt of the joke, the less amusing it is. Moreover, a loft to the ego does not always cause laughter. Indeed, we often laugh in sympathy with others.

> *He is a writer for the ages—the ages of four to eight.*
> DOROTHY PARKER

The Parker witticism has two parts—the writer is eternal, the writer is infantile—and they seem to merge, yet clash, before a final meaning emerges. They are incongruous. Incongruity theory holds that laughter arises when two or more inconsistent ideas try to meld. Numerous thinkers have noticed this feature. In 1777 Joseph Priestley wrote that laughter stems from perception of contrast. In his *Critique of Pure Reason* (1781) Immanuel Kant said it arises "from the sudden transformation of a strained expectation into nothing"—also the formula for dashed hopes. More moderately, Herbert Spencer said the incongruity must merely descend: We gear up for the great and encounter the small. (Movement from small to great, he said, causes wonder, though wonder itself can cause laughter.)

Wit is "the disguised priest who marries all couples," declared Jean Paul (1763–1825). We laugh at the petty, since it contrasts with our deep sense of the sublime. A jest at the expense of others gives them the prize of our insight, he felt, so "the poetic bloom of its nettle does not sting."

In one celebrated twist on incongruity, Henri Bergson held that we laugh at the mechanical foisted on the living. A person acting like a robot—a mime, say—is funny. More broadly, a situation is always comic when we can give it two completely different meanings at the same time.

Though some of these approaches seem narrow, incongruity is plainly important in humor. One recent study of 242 Chinese jokes found that 210 of them involved it. And yet 32 didn't.

> *Having been forsaken by Dame Luck, he*
> *degenerated into a Lame Duck.*
> ANONYMOUS

This line ends with a jolt, a sudden play on words. Surprise mounts the saddle in the third theory. Many other thinkers invoke it; for instance, it is the "sudden" in Hobbes's "sudden glory." But theorists of this stripe hold it is the crux. Descartes said that laughter arises from the mix of shock and moderate joy. In 1940 psychologist John Willmann declared that humor stems from situations inducing both surprise/alarm and playfulness. Surprise theory explains why most jokes aren't funny the second time around. It labors to explain why a film like *A Fish Called Wanda* can be funny on second viewing, why we can laugh in recalling an amusing line or event, or why a waddling penguin is comic, though we've seen it endlessly before. When comedian Andy Kaufman walked onstage and began reading from *The Great Gatsby*, the audience eventually started laughing, perhaps realizing there would be no punchline. The theory is silent before uneasy laughter, which is rarely sudden.

> *One would have to have a heart of stone not to*
> *laugh at the death of Little Nell.*
> OSCAR WILDE,
> on *The Old Curiosity Shop*

Ambivalence theory maintains that we laugh when we feel incompatible emotions at once. While incongruity treats ideas in the joke itself, ambivalence focuses on our reaction. In 1560 physician Laurent Joubert (1529–1582) issued *Treatise on Laughter*, the first modern book on the topic. In it, he claimed that sorrow shrinks the heart while joy inflates it. Felt together, they make it rapidly throb, shaking the lungs and causing laughter. This mechanism is now an intellectual curio, but the larger theory lives on. Baudelaire said, "Laughter is satanic; it is thus profoundly human," and added that it "is at once a token of an infinite grandeur and an infinite misery." Other writers have suggested that laughter bubbles up from the clash of love and hate, play and sobriety, sympathy and animosity, mania and depression.

In 1983 John Morreall said the cause of laughter is simple: an abrupt, pleasant psychological shift. A catapult to superiority qualifies, for instance. When we see an old friend, we go from neutrality to pleasure and hence laugh. What about embarrassed laughter? Morreall argued that we first feign it, but since even feigned laughter is pleasurable, we then break out in real laughter. The last example wobbles, and in 1994 another scholar countered with six kinds of laughter he said flout the theory: hollow or mirthless, nervous, imitative or contagious, hysterical, tactile (from tickling), and pathological.

Indeed, some claim that feeling, far from being the source of humor, is the antithesis. Horace Walpole famously wrote, "Life is a comedy to the man who thinks, and a tragedy to the man who feels," and Bergson thought emotion was laughter's nemesis.

> *A: "Kind? He remits nothing to his family while he's*
> *away. Do you call that kindness?"*
> *B: "Yes. Unremitting kindness."*
>
> DOUGLAS JERROLD

Here, B twists an earnest accusation into nonsense, dissolving it. The joke yields relief. Relief is implicit in all these approaches, but strides to the fore in release theory. Herbert Spencer viewed laughter as a discharge of excess nervous energy, which has no other outlet. Dewey said laughter marks the sharp end of a period of suspense. Theorists like J. C. Gregory contend that relief is the root of laughter, from which its many varieties arise. In fact, we do laugh out of pure nervousness, when we are not at all amused. Moreover, tension makes people laugh more readily and heartily. As Darwin noted, soldiers just back from mortal danger guffaw at the slightest cuip. Audiences at suspense films are also easier to amuse — one secret of "buddy" movies like *48 HRS.* But this theory, alone, doesn't explain the cause of laughter.

> *The botanical gardens boast many varieties of cactus*
> *not found anywhere, not even in the botanical gardens.*
>
> S. J. PERELMAN

Configuration theory holds that we laugh when apparently unrelated elements fall into place all at once. We perceive meaning in a gap. It's like peering through a keyhole and seeing orchids. We experience abrupt insight, and this pleasing process makes us laugh. Hence the twice-told jest isn't funny because no insight occurs the second time. Similarly, a joke can be too obvious, so insight is unnecessary, or too remote, so it is impossible. This theory explains why timing is crucial for comedians, since it affects the inference process. It explains why brevity is the soul of wit: It creates omissions. Exegesis of a joke fills in the gaps and is famously dreary. The theory also suggests why jokes establish bonds: They show that two people think alike. As Goethe said, "There is nothing in which people more betray their character than in what they laugh at."

> *Peccavi. ("I have sinned.")*
> CATHERINE WINKWORTH (dispatch
> supposedly sent by Sir Charles Napier on capturing the Sind
> in 1843)

Economizing theory holds that laughter saves on mental processes. Its foremost figure is Sigmund Freud, whose *Wit and Its Relation to the Unconscious* (1905) remains the most elegant survey of this ground. Freud worked from concrete examples — actual jokes — and stated the problem clearly, so his prose sparks ideas even if one discards his conclusions. He anatomized laughter into some twenty areas, each with its own laws, but his three umbrella fields are: wit, the comic, and humor. Wit includes jokes and nonsense, and requires a human inventor. We laugh at it because it economizes on *inhibitions*, that is, it briefly diverts the superego so we can release our repressions. The comic involves unintentional discovery and includes waddling penguins, mishaps on *America's Funniest Home Videos*, and *Plan 9 from Outer Space*, as well as parody and caricature, which render people comic. "Wit is made, while the comical is found," he says, momentarily forgetting parody. The comic economizes on *thought*, since it spares us the effort of understanding others. Humor saves on *feeling*, turning an event that might spur suffering into one that doesn't, as in the works of Molière or Twain.

★   ★   ★

Despite the jokes introducing the theories above, each theory seems relevant to every joke. In fact, all the theories are pertinent and none suffice. Like flashbulbs in the dark around an exotic artwork, each illumines a different part of the puzzle. Superiority and incongruity focus on the content of jokes, surprise and ambivalence on our emotions; configuration on the process of insight, and release on the upshot. Freud's multitiered approach embraces them all, but ties them into psychoanalysis, unproved and perhaps unprovable by science.

Willmann said the main obstacle to a coherent theory was the variety of laughter itself. Then he uttered the fatal words, "There must, however, be some basic principle," and plunged ahead. Freud shunned this formula hunt and offered a basketful of explanations. He also denied his theories were comprehensive, and with good reason.

Laughter has so many sides, and we laugh at myriad phenomena. In 1902 James Sully listed twelve: 1) novelty, 2) physical deformity, 3) moral weakness, 4) disorderliness, 5) minor misfortune, 6) indecency, 7) pretense, 8) ignorance or lack of ability, 9) incongruity or absurdity, 10) word play, 11) besting another, and 12) sheer good spirits. This inventory is clearly incomplete. Other, nonhumorous sources of laughter are tickling, nervousness, play, relief, triumph, setback, confusion at an utterance, and, arguably, nitrous oxide. One despairing researcher dismissed all such lists as futile "since man apparently laughs at just about everything."

Beyond the general theories lie other issues that have attracted attention.

Which comes first, the pleasure or the laugh? Most believe we feel pleasure and laugh to express it. But William McDougall demurred, saying we laugh because we feel miserable, and the act makes us feel good. "The perfectly happy man does not laugh," the psychologist stated, "for he has no need of laughter." Beaumarchais said, "I laugh at everything, for fear of being obliged to weep," and Nietzsche added, "Man alone suffers so excruciatingly in the world that he was compelled to invent laughter." In fact, misery can trigger laughter. But this theory does not explain The Marx Brothers.

There are other questions, and the chatty Joubert, in his semiprofane *Treatise*, raised a welter. Can we laugh in our sleep? (Sure, he says. Why not?) Why does wine make some people laugh and others cry? (It depends on their natures, and whether the wine is good.) How come we occasionally can't suppress our laughter? (Don't worry about it, he counsels. "Laughter is voluntary, whether you want it to be or not.") Why do great laughers grow obese? (Laughter vaporizes the blood, which spreads through the body and produces fat.) Can laughter cure the sick? (It helps, and he recommends monkey-watching. Norman Cousins popularized the idea in his *Anatomy of an Illness* [1979], claiming ten minutes of belly laughter gave him at least two hours of painless sleep, and scientists have since confirmed that comedy kills pain. So does tragedy.)

What causes laughter? Can mortals define it at all, or does it lurk forever in some black Gödelian limbo? One professor recently claimed a link between humor and "chaos theory" — not Lorenz equations or Mandelbrot sets, but chaos in the popular sense as complexity beyond our power to codify. Humor, he wrote, is a chaos we create for our own entertainment.

Chaos or not, humor research is burgeoning. The number of academic papers on humor tripled between 1970 and 1990. The first international conference on humor occurred in Wales in 1976, and several others have followed. The academic journal *Humor* appeared in 1987. More scholars are probing this mystery today than at any time in the past.

And most know the pitfalls. The subject is everywhere and evanescent, and the literature quite humorless. Laughter is a sprite that must be dissected alive and fully conscious, when it can easily flee the scalpel. Bergson in 1911 compared laughter to the foam on a wavelet. "Like froth, it sparkles. It is gaiety itself. But the philosopher who gathers a handful to taste may find that the substance is scanty and the after-taste bitter." So far, laughter has had the last laugh.

The Merry Hominid

George Eliot wrote that humor probably originated in "the cruel mockery of a savage at the writhing of a suffering enemy — such is the tendency of things toward the better and more beautiful." But evolution is thrifty, and humor probably did not arise for this unimportant job. Why did it? What's the point of laughter?

It's a more serious question than most people realize. We tend to view laughter as a frivolity, but it is ubiquitous and often intense. It shapes social gatherings, friendship patterns, even mating. And it's deeply enjoyable. Jules Renard (1864–1910) said, "We are in the world to laugh," and people who dislike laughing are rarer than those who dislike life itself.

Such facts imply a major evolutionary role, since all facial displays are costly. They consume energy and can attract predators — and laughter is especially noisy — so they must pay their way. Normally they influence others to one's advantage. We've seen how human sadness, pain, and other expressions move people. What about laughter?

As with smiles, the natural world offers clues. Aristotle said we are the only animal that laughs, and technically he is correct. But other creatures don't laugh for much the same reason they don't speak: they lack the equipment. In laughter, we repeatedly block the outflow of air from the lungs, usually with glottal stops or glottal fricatives. (There is a glottal stop before "ice" in "an ice man," which distinguishes it from "a nice man.") Unlike Arabic, say, English has no character for the glottal, so in laughter we denote it with a string of *h*'s. Almost any vowel can appear in between, but we usually stay with one, as in *ha ha ha* and *ho ho ho*. Diphthongs — like the lewd *hyeugh hyeugh hyeugh* of Pig Bodine in *V.* (1963) or Renfield's creepy *hyooong hyooong* in the original *Dracula* (1931) — are uncommon, and a whispered laugh is also rare, usually an upshot of stealth.

Animals don't laugh like we do, but do they laugh at all? Darwin seemed to think so. Tickle a young orangutan, he noted, and it grins and makes a chuckling sound. Its eyes brighten and when the tickling stops, a smile-like expression illumines its face. At the London Zoological Gardens, he observed the Barbary ape move its lower jaw up and down, bare its teeth, and wrinkle its eyes, producing a sound "hardly more distinct than that which we sometimes call silent laughter." Keepers told him that it was the ape's chuckle. Observers have recorded laugh-like utterances in gorillas, baboons, and macaques, and Jane Goodall believes chimps laugh when wrestling or tickling.

Based on such findings, psychologist Glenn Weisfeld of Wayne State University has advanced a tentative theory of the evolution of laughter. He suggests it began with tickling. As Darwin noted, we are ticklish in areas others rarely touch, like the ribs and soles of the feet. In children tickling seems like a play-attack. It helps them develop defensive reflexes in these areas, and their giggling encourages the adult to persist. As we grow older and master these reactions, we can come to find tickling unpleasant.

From tickling, laughter generalized to play-fighting. Primates, children, and young adults often laugh when hitting or wrestling each other in fun. Their laughter marks the combat as a game and protects against damage from misunderstanding. Mock battles, of course, enhance real skills.

And from tickling and physical play, Weisfeld suggests, laughter became a response to social play, where it also encourages practice. Banter, for instance, is a play-attack and sharpens our ability to react to genuine insults. Jokes demand inference, a vital skill everywhere. Play with incongruity invites resolution of mixed messages, which can signal trouble in the real world. Even word play, he suggests, can seem amusing to adults because it hones the language skills we develop throughout our lives.

Weisfeld is cautious with these ideas, and sensibly so, for laughter seems to have ramified like twigs on a bush, beyond any neat, linear description.

For instance, in one surprising study, Robert Provine of the University of Maryland showed that over 80 percent of laughs in normal conversation had nothing to do with humor. They occurred after statements like "I'll see you guys later!" and "I know!" and questions like "How are you?" and "It wasn't you?" He also found that speakers, especially women, laugh more than listeners. Most such laughs occur in playful settings, and good mood spurs them more than any words. The bubbly sound of party mirth stems from this kind of laughter, not from the wit of guests. It is the paralanguage of laughter, both a punctuation of speech and a token of playfulness and good will.

Much of the remaining 20 percent of laughter is genuinely humorous, and it too serves as social glue. As such, it offers two rewards: cooperation and power.

Laughing clearly furthers cooperation, that mammoth boon to our species. Humor is a weld of pleasure, and we feel closer to those we laugh with. Thus one study found that mutual ribbing in Chippewa tribal councils enhanced solidarity. Laughing pre-humans likely enjoyed one another better, hence worked together more easily.

They may also have understood and trusted one another better. Indeed, humor meshes minds. It demands insight and social sense, a feel for the values of the audience: whom and what the listeners respect, like, resent, disdain. As advocates of configuration theory note, wit and laugher must share similar assumptions, fill in the same way, and thus humor can quickly reveal bonds between strangers.

Humor is also a tool of tact. It can smooth a looming offense into play, suggesting goodwill while discreetly shifting the problem into limbo. If one person puts another in an uncomfortable position, as by urging her to buy an unwanted item, she can escape with a joke and both will remain friends. The noose becomes a slip knot.

And perhaps most basically, laughter rewards us for simply being together. For instance, others' laughter intensifies our own. It turns the bare risibility into a chuckle and the open-mouth laugh into a roar. We laugh harder in a packed moviehouse than in an empty living room. This phenomenon has attracted much attention and we still don't understand it— though smiles, disgust, and sadness are contagious too.

The ability to gain cooperation amounts to social power. Indeed, we compete for the alliance or allegiance of others, in the game called politics, and dominance and prestige wreathe its winners. Success here advances people in the less obvious com-petition to pass on their genetic blueprints, and men especially benefit. Most human societies have been polygynous, and pow-erful, high-status males have had multiple wives, committed more adulteries, and left behind more DNA. Humor helps loft one to this evolutionarily desirable status.

It works politically in many ways. For instance, humor is persuasive. As Freud noted, a joke on behalf of a cause inclines us to it, and hence presidential candidates often hire gagwriters. Humor can enforce conventional values, as in Restoration comedy and Augustan satire. It can forge in-group coherence, exalting one's own set and belittling rivals.

Humor also determines status more directly, by rewarding friends and scalding foes. We don't bite each other like chimps, but a clever jest can be more damaging, since it reduces the victim to an object of play, to social insignificance. Hence Greenland Inuit often resolved disputes with public contests of ridicule, and youths in U.S. street gangs engage in competitive persiflage in which they try to "rank" the opponent. One study of hospital workers showed that mockery traveled downward through the social hierarchy.

In addition, humor fosters freedom. Jokes often garb aggression in play, making it more acceptable. A sense of humor confers the right to poke and thrust, under the forgiving nod of fun. As David Letterman says, after skillfully drawing blood, "It's just a *joke!*" Humor also frees one to speak of charged and somewhat taboo topics, like sex, violence, and major controversies, again by making them amusing. Indeed, by leaving a gap at the right place the jokester can make the audience supply the taboo thought. It all expands one's social power.

Scientists know that humorous people react better to stress. Laughter itself reduces both physiological and psychological stress, as Spencer's release theory predicts. Psychologists like Rollo May have speculated that the amusing can better put their problems at a distance, gain perspective on them. But they may simply have more control socially, more power over stress-causing events. Researchers have found that humor corresponds to positive self-image, and there are many reasons why it would.

Not surprisingly, people choose mates partly based on humor and laughter, which play a direct role in courtship. In one intriguing study of conversations among young German strangers, the more intensely a man made a woman laugh, the more she wanted to see him again. But a woman's wit did not affect men the same way. Rather, the more a woman laughed at a man's quips, the more he wanted to see her. Like enlarged pupils, women's laughter indicates interest, and indeed women know the strategic importance of laughing at men's jokes. Men, on the other hand, know the importance of putting a sensitive advance in a context of play.

Overall, these tactics make genetic sense, since people are favoring mates who are likely to cooperate well and hence have more friends and political clout. The pair will probably understand each other better, thus stay together longer to raise children. The husband is more likely to be able to protect and provide for the family, and the wife to be loving with children. And the male, indeed, may pass on genes with more than one wife.

Scholar Geoffrey Miller goes further and suggests that humor and kindred traits actually caused the great growth of the hominid brain. He cites surveys that show we commonly list creativity, sense of humor, and interest of personality above even wealth and beauty in the ranks of qualities we desire in the opposite sex. Hence, he asserts, the brain is mainly a courtship device, like the peacock's tail. Prehumans sought out "psychologically brilliant, fascinating, articulate, entertaining companions," selecting each other for humor, musical and artistic talent, and general creativity. If so, Renard is literally right. We are in the world to laugh.

Blushing

Why do we blush? "It makes the blusher to suffer and the beholder uncomfortable," said Darwin, "without being of the least service to them." It is unwanted and can get worse just because we realize it's happening. It is not muscular, like most facial expressions, but vascular, a sign from the blood. Other animals have the physical equipment, and monkeys flush red in passion, but, as Mark Twain said, "Man is the only animal that blushes. Or needs to."

Though Darwin described blushing in detail, it has attracted surprisingly little attention since. The psychoanalysts offered an astonishing — and blush-causing — explanation for it: It is exhibitionism by proxy. The blusher yearns to display the genitals and uses the face as a substitute. Existentialists like Sartre felt blushing showed awareness of one's body as others saw it. Existential analyst Ludwig Binswangler said blushing stemmed from a touch on the "inner border of sin," involuntarily revealing its locale.

Annoyingly, blushes love an audience. Darwin blushed in solitude and some people crimson on the phone, especially after obscene calls, but they are unusual. And for most people, onlooker reaction is far worse than the burning glow itself. Over 81 percent said observers comment on a blush. Adolescents in particular find people call a blush to their attention.

Even the stages of blushing seem a snide prank. Blushes occur quickly, usually within two seconds of provocation. The cheeks redden before we feel the heat, so others know we're blushing before we do. Sometimes we don't realize it at all. In one study, over half the subjects said people had told them they were blushing when they didn't feel it.

Blushing goes with confusion of mind — distress, stammer, averted eyes, awkward movements — partly because it flings open a window to secret parts of the mind. It strips away privacy, often when we want it most. In *The Princess of Clèves* (1678), the princess blushes often, unveiling her passion to her yearning lover and her guilt about it to her husband. In one experiment, a thirty-year-old man said, "Five months ago I was with three people, [and] one was an ex-girlfriend. One of the friends asked me if I had had an affair over the summer, and my answer was no to the specific question, but in actuality the reply should have been yes. After the blush appeared the subject was changed very quickly."

In another study, interviewees mentioned three kinds of spur: 1) another's remark or act, such as "I was at my boyfriend's house and his cousin asked if we were going to get married, in front of all those people" (55 percent of cases); 2) one's own thoughts or faux pas, such as, "I accidentally splashed food on a woman while serving dinner" (15 percent); and 3) other situations, including being made socially conspicuous, such as, "I was asked out by a great-looking guy who I thought liked my friend — so I was really shocked. I looked down when I felt myself blushing" (24 percent).

We blush more readily before the opposite sex. Darwin thought women blush more often than men, and in his day they may have. Studies today show little difference, even though women overall are more emotionally responsive than men, especially in the face.

Some people blush once a day or more. Chronic blushers can fear going places where they might meet other people. Darwin wrote, "Some persons, however, are so sensitive, that the mere act of speaking to almost any one is sufficient to rouse their self-consciousness, and a slight blush is the result."

Blushing follows a clear trajectory through life. It begins as early as three, becomes most common and intense in adolescence, then wanes, especially after thirty-five. Investigators have suggested many reasons for the peak at adolescence, such as "self-consciousness and embarrassment following rapid bodily changes; hormonal changes; unsettled identity, particularly sexual identity; new kinds of social encounters which provide the setting for interpersonal judgments." Adults, in contrast, often have "well-defined and well-rehearsed" roles in which they are competent, and more status in general.

Blushing is universal. Blacks blush, but it appears as deepened blackness, harder to perceive. One 1990 study found that two-thirds of people feel blushes mainly on the cheeks, while a quarter feel them spread over the whole face. In addition, 26 percent said they blushed on the ears, 21 percent on the neck or chest, and 6 percent on the scalp, the classic "blush to the roots of the hair." Darwin wondered why blushes don't spread over the whole body, and suggested it's because the face commands our attention. Blushing is a signal, and the face is the body's best relay point.

What does blushing convey? It isn't anxiety or fear; anxiety actually blanches the face. Many observers have suggested shame or guilt. Macrobius (5th cen.) said, "Nature being moved by shame spreads the blood before herself as a veil." Hobbes noted that shame "discovereth itself in blushing," and according to Rousseau, "Whoever blushes is already guilty; true innocence is ashamed of nothing." In 1839 Thomas Burgess declared blushes show an inborn sense of guilt in straying from God's path.

But embarrassment comes closer. Exposure of body parts, betrayal of sexual thoughts, and lapses of etiquette all trigger blushing. Overpraise can cause embarrassment and facial reddening too, if one feels one doesn't live up to it. And since a blush per se can be embarrassing, it can cause further blushing.

Yet a blush is not quite the insignia of embarrassment. Some embarrassments, like performing poorly on a test, may not spur blushing. And in turn, nonembarrassing experiences can cause it. People blush when receiving awards or speaking before groups. The center of attention often crimsons when others sing "Happy Birthday." At a ball in *War and Peace*, Prince Andrei tells Natasha he had inadvertently overheard her on an earlier moonlit night. "Natasha blushed at this reminiscence and tried to excuse herself, as if there had been something to be ashamed of in what Prince Andrei had accidentally listened to." She doesn't feel embarrassment, but exposure. The remark makes her self-conscious.

Public self-awareness is critical. Darwin felt the states that induced blushing were shyness, shame, and modesty, all characterized by self-attention. It is the "thinking what others think of us, which excites a blush." Thus people can blush when praised. And we don't blush at the thought of our actual guilt, he said, but rather at "the thought that others think or know us to be guilty."

Why should we have to reveal embarrassment and self-consciousness to others? Why did the blushing gene spread? Darwin thought blushing conferred no advantage and deemed it a relic of some prior function.

But it does have use. Blushing requires sensitivity to other people, who can find it charming. During the engagement between Natasha and Prince Andrei, she asks about his son. "Prince Andrei blushed, as he often did now — Natasha particularly liked it in him — and replied that his son would not live with them." A blush shows concern for the opinion of others, which may explain their delight in pointing it out.

Darwin claimed blushing was not a "sexual ornament," but he may be wrong. Blushes can play cupid's aide, revealing affection and desire. By revealing the importance one person attaches to another's opinion, it can encourage romance. This fact may partly explain its prevalence in adolescence.

And in folk wisdom, blushing is a guarantor of honesty. Members of the Hagen cargo cult in Papua New Guinea speak of "shame on the skin," which means one has a soul, a good social attitude. When a person doesn't blush, a wild spirit that lives by riverbanks has lured his soul away, driving him crazy. Diogenes the Cynic (4th cen. B.C.) said, "Blushing is the color of virtue." "Better a red face than a black heart," runs a Portuguese proverb, and poet Edward Young (1683–1765) wrote, "The man that blushes is not quite a brute."

The Fifth Amendment does not apply to the face. It commonly incriminates us, and to the extent blushing reveals our discomfiture to the group, it may foster cooperation. As primatologist Frans de Waal observes, blushing may show that we have internalized moral rules and are trustworthy, hence boosting our chances of survival. Its involuntary nature makes the signal reliable, indeed, possible, since few people would ever issue it on their own.

Most explanations of blushing, however, have ignored its role among blacks, even though it almost certainly evolved in Africa, when we were all black. In one recent study 30 percent of blacks said "no one ever notices [my blushing]" and just 22 percent said others see a blush through change in skin color. At the same time, 59 percent of blacks said other people detect their blushing because "I act embarrassed," compared to 25 percent of whites. Among blacks, blushing is not a bold facial flare, but more the remaining behavior: confusion, awkwardness, a sense of lost privacy. Hence Darwin may be partly right. The crimson face may be a relic, an exaggeration of a signal meant to be subtler.

Blushing remains curiously elusive, shot through with puzzles, such as why it should be so intense after breaches of etiquette. Despite recent work, many scientists still view it as a kind of Van Diemen's Land, half-charted and waiting.

The Power of Staring

At Tel Brak, an excavation in eastern Syria, archeologists have uncovered a temple housing thousands of alabaster figurines. They all have the same shape: two bulging, alligator-like eyes atop a short neck and a torso. Archeologist M. E. L. Mallowan has dated them to 3000 B.C. They are pure stare.

Staring is special. Our mental gaze-radar detects it quickly, and even if we are utterly safe, we feel a quiver of warning inside. The omen is visceral, deep, oddly maddening. And it's not just psychological. Being stared at increases arousal, raising the heartbeat and altering the galvanic skin response, especially if the victim cannot counter or escape.

This kind of autonomic reaction forces us to deal with the starer. It forges a social tie out of vapor. Hence beggars stare in solicitation. Manet shocked the art world with *Olympia* (1863) and *Le déjeuner sur l'herbe* (1863) in which naked women gaze straight at the observer. Before, nudes had looked discreetly away, allowing perusal of their bodies. These stares turned the observer from voyeur to participant, and it was too much for many Victorians.

When we can't halt a stare, we tend to flee. In one study, researchers on a street corner gazed steadily at drivers who had stopped at a red light. The drivers noticed the stare within seconds and when the light turned green, spurted across the intersection significantly faster than other drivers. Staring at pedestrians makes them walk faster, and students stared at in a college library leave earlier.

Sensitivity to stares is genetically ancient. It goes back to reptiles and insects. The hog-nosed snake will fake death longer if a predator is staring at it than staring away. Lizards too freeze longer in the presence of a person gazing at them. Plovers with eggs in the nest move off it in decoy and stay away longer when humans stare at them. Such evidence suggests staring is the oldest and most stable facial expression on earth.

And the reaction to it seems innate. Monkeys reared in total isolation try to conciliate a staring, full-face picture more often than a profile. In one remarkable study, baby chicks shunned a darkened circle more than a rectangle, and a pair of circles more than one. And if the twin disks seemed to follow them, the chicks stayed even further away.

Even at the subtlest level, the gaze may be a disruption. For Sartre, it symbolized the intrusive presence of another. Alone in a park, he said, we survey the lawns, trees, and brooks, and we are the center and reigning consciousness of the world. But if another person enters, the very nature of the park seems to change. A second force field appears, and we go from watcher to watched, emperor to environment, willy-nilly.

Some stares externalize conscience. The eye peers into us and sees our awful secrets. Man Ray (Emanuel Rabinovich, 1890–1976) placed an eye in a metronome. Dr. T. J. Eckleberg's huge eyes gaze down from a billboard in *The Great Gatsby*, unnerving sinful drivers. In Edith Wharton's (1862–1937) short story "The Eyes," the narrator sees a pair of hideous orbs twice in his life, once when he is about to propose to a woman he doesn't love, and again after he encourages an inept young novelist, out of friendship. They appear at night, at the foot of his bed, and express a confident turpitude, built up "coral-wise" from subtle offenses over decades. In old age, he realizes the eyes were his own, which had grown to match them.

A hard gaze has aided the career of many an actor. Audrey Meadows's level stare in *The Honeymooners* was the ultimate weapon in Kramden spats. Paul Newman's eyes are famous partly for their cool, direct gaze. George M. Cohan said Spencer Tracy could "stare, glare, and finally scare the other actors," and this power fitted him for his many roles as walking conscience.

Staring feels like mind-burrowing, and sometimes we relish it. Lovers gaze at each other and feel entwined. In *Antony and
Cleopatra*, the vampish queen says, "Eternity was in our lips and
eyes." The flirtatious look welcomed by both parties is a visual
intimacy, and invites physical and emotional approach.

But staring is usually rude. It invades others, psychologically and physiologically. The classic stare is blatant, persistent,
often blank, and unresponsive to the acts of its target. For
instance, the flirtatious glance is cautious and vulnerable, but the
leer is the opposite, a burning probe.

To avoid rudeness, we normally don't gaze straight into
people's eyes when we speak to them. Rather, our eyes dance
about their faces or even turn slightly aside, and in fact men
look less at the face than women. When we tire, however, we
may stare, a sign that our social systems are flagging. Staring can
indicate other kinds of flawed sociality. For instance, schizo-
phrenics stare, perhaps in wonder at a senseless world.

We stare at objects of great interest, like movie stars. Dennis
Hopper says when people gaze at him in a shoe store, he first feels
they think he's a shoplifter. Then he says to himself: *Ah, they're
looking at Dennis Hopper.* Other celebrities, like Winona Ryder,
find the endless public gawk more upsetting.

We stare at physical abnormalities, as the disabled know so
well. Yet the impulse feels ugly and makes us notably uncomfortable. Intriguingly, we also gaze at pregnant women. In both
cases we strive to hide our interest, letting our eyes linger mainly
when we are unseen or in conversation. One team of scientists
speculates we stare to assimilate the novelty. We wouldn't gape
at pregnant women in an obstetrician's waiting room, say,
where they are common.

A step beyond rudeness, staring becomes a blunt power
device. Among chimps, gorillas, and a wide range of monkeys,
males use it to control subordinates. For instance, a macaque first

glares at his inferior. If the other monkey ignores this signal, the male makes a more serious threat, opening his mouth into a gape-stare. The recipient can turn away and submit, or defy the warning and stare back, in which case the alpha-male attacks. The signal may be inborn. Infant monkeys raised in solitude make more appeasement gestures to a photo of a staring face than a profile.

Staring precedes attack in humans as well, and retains the aura of threat. Indeed, the drill-like gaze emblemizes personal power. In the Mesopotamian *Epic of Creation*, godlike Marduk has a proud form and piercing stare. Rasputin and the Ayatollah Khomeini had knifelike eyes, and mystic and con man George Gurdjieff deployed a memorable gaze. In 1921 Gurdjieff persuaded two landlords to sign all their leases over to him and, when tenants sued, the duo claimed his eyes had hypnotized them. It was a common complaint through his career.

One critic has discussed Picasso's portraits in terms of the Andalusian notion of *mirada fuerte*, or "strong gazing." *Mirada fuerte* is a way of dominating others, and Picasso, notorious for his need to control people, used the face-forward pose almost solely in self-portraits. In portraits, his subjects gaze off to the side, as if unable to tolerate his powerful eyes.

Looking away from a speaker can convey boredom or embarrassment. But in response to the challenge of a stare, it can signal shyness, guilt, or weakness. Wolves turn away from the gaze of a superior, and in one study where an interviewer negatively evaluated subjects, they avoided looking at him. Hence staring smacks of a contest of strength, a notion that lies behind terms like "face off," "outface" (to cause to waver or submit as if by staring), and "face down."

The gaze of an image can have surprising power, as with *Olympia* and Dr. Eckleberg. In one interesting study, researchers aimed a female bust at overweight women in a college cafeteria.

They departed faster than overweight women the object did not target. Even schematic eyes draw our attention. Marquesan warriors tattooed concentric circles called *ipu* under their arms, so when they raised a club in battle, the *ipu* flashed out and startled the enemy.

Hence the value of staring posters to tyrants. Assad and Hussein gain an edge from their inescapable gaze, which reminds citizens that their rulers are everywhere and can peer into their disloyal souls. Indeed the symbol of Hussein's secret police is an eye on a map of Iraq. Fictional power icons are also ubiquitous starers, and Winston's prime concern in *Nineteen Eighty-Four* is finding privacy. He lives in a nightmare of eyes.

Ultimately, the stare is deadly — in folklore, at least. The Irish giant Balor struck men dead with his glance. The torpid Serbian monster Vy lay all day on an iron couch, blinded by the weight of his eyelids, but when danger arose, he called in twenty heroes who lifted his lids so he could rake the world with his killing stare. Banned from teaching, Rabbi Elieser Ben Hyrcanus scorched the ground with his gaze and made the pillars of the scholars' meeting place tremble. We have no idea if they reinstated him. Mandeville tells of an isle in the Indian Ocean of "cruel and wicked" women with gems growing in their eyes. Their angry look at a man slays him. The lethal-eyed Scorpionmen in *The Epic of Gilgamesh* guard the sun at dawn and dusk, when their glowing mantles drape the mountains.

The glance of the Medusa petrified humans. The legend has many versions, but in the canonical one, a king asks Perseus to acquire the head of Medusa, the only mortal among the three Gorgons. Athena and Hermes give him the helmet of invisibility, a pouch for hiding the Medusa's head, and winged sandals for fleeing the sisters. He reaches the shores of Oceanus and finds the trio asleep. Using a mirror on Athena's shield to view them, he decapitates Medusa and tosses her head in his pouch. The Gorgons pursue him, shrieking wildly, but he dons the helmet of invisibility and escapes.

Even dead, the Medusa's head could turn living things to stone, so its owner became a superman. With it, Perseus freed Andromeda from the sea monster Cetus and later petrified the king who'd sent him on the mission. Ultimately he presented the head to Athena, who mounted it on her shield.

The Medusa was a popular image in the Greek world and appeared throughout it, from Sicily to southern Russia. She is very appealing, with her round face, flat nose, tusks or fangs, and tongue stuck out to her chin. The fatal head adorned city walls and the breastplates of soldiers, to terrify the foe.

The basilisk was the Medusa of the animal world. According to Pliny, citizens who glanced at this foot-long serpent died at once. Its breath could crack rocks and once, when a rider on horseback killed one with a spear, its fumes slew both man and steed. The real basilisk (*Basiliscis basiliscus*) is a harmless Central American lizard with its own semimagic power: It runs upright on water, supported by tiny air pockets beneath its toes.

The lethal stare makes for interesting legends, but who has ever seen one? Its very power dissolves its credibility, and leaves the human mind vulnerable to its sly, diffuse cousin.

The Evil Eye

In *Anna and the King of Siam*, the true story of Anna Leonowens and the ultimate source of *The King and I*, a British diplomat arrives at the Thai court on vital business. To honor him, harem women politely greet him in European hoopskirts. But his full beard shocks them, and when he dons a monocle, they think it is the evil eye and flip the hoopskirts up over their faces.

*Jettatore!"** as he passes. Shopkeepers stick their first and fourth fingers out at him, and women shake their fists in his face. "What is there strange, peculiar, or ridiculous about me that attracts such unpleasant attention?" he wonders. Misadventures compound. In the end his fiancée dies and he leaps off a cliff.

King Alphonso XIII of Spain (1886–1941) developed a reputation as a *iettatore* after his state visit to Italy in 1923. As his ship approached Genoa, the sky was clear, but suddenly a great gale arose and swept four sailors to their deaths. An air compressor in a nearby Italian submarine blew up, killing a man. At public appearances, Alphonso now heard the prophylactic jangling of horseshoes and keys. The calamities continued. As the king entered the Bay of Naples, officials fired an old bronze cannon in salute. It exploded, killing the crew. A naval officer in the reception line collapsed after shaking hands with Alphonso and died in the hospital. At the end of the visit, the king passed the dam holding Lake Gleno. On the following day the dam broke, killing fifty and rendering five hundred homeless.

The Italians did not forget Alphonso. Exiled in 1931, he visited Italy seeking to finagle a way back onto the throne. Mussolini refused to see him, and nervous servants jingled iron keys in their pockets as they waited on him. Elsa Maxwell gave a party at which only Alphonso showed up; twenty Italians had other obligations. Eventually Alphonso took to entering moviehouses after the lights dimmed, to avoid notice.

Pope Pius IX (1792–1878) also had the evil eye. Italians first realized it after he gazed at a window where a nurse stood with a child in her arms. Minutes later the babe fell to earth and died. According to American sculptor and writer William Story,

---

*The *j* has almost vanished from Italian, and today "evil eye" is *iettatora*, and "one who casts the evil eye" is a *iettatore*.

one Roman said, "If he have not the *jettatura*, it is very odd that everything he blesses makes fiasco. We all did very well in the campaign of '48 against the Austrians. We were winning battle after battle, and all was gaiety and hope, when suddenly he blesses the cause, and everything goes to the bad at once." His successor, Leo XIII (1810–1903), also gained this reputation, after a large number of cardinals died during his tenure.

Intriguingly, one can cast a spell on oneself. Romanian children under one year old can give themselves the evil eye, so parents shield them from mirrors. An especially vile *iettatore* in Messina accidentally killed himself in 1883, rumor held, by glancing at himself in a mirror as he walked down Corso Garibaldi. In *The Idylls* of Theocritus (c. 270 B.C.), the herdsman Damoitas admires his face in the sea, then worries about this fate too, so he spits three times on his breast, as a crone had taught him.

Prophylactics against the effects of the evil eye are global and diverse. A common one is wearing red, and Rumanians will tie strips of red cloth to animals. According to the U.S. Defense Department, Manuel Noriega wore red underwear for this purpose. Turks favor blue fabric. In India, some people use black cosmetics to ward off the evil eye at marriages and funerals. South Indian women put lampblack on their eyelids to avoid casting or succumbing to the evil eye. The Koravas, a caste of thieves, employ special soldering of the sharp end of the tool they use for break-ins.

Italians guard against it with horn amulets, garlic, outstretched fingers, forked coral, pig's teeth, spitting after the *iettatore*, and uttering verbal charms, such as: "Away, away, away! Tuna eggs in France, Let bad luck go to sea!" But the supreme defense is metal. Prudent citizens nail horseshoes to the walls of homes and touch keys in their pockets when they meet a *ietta-tore*. After the murder of Joseph Buquet in *The Phantom of the Opera*, the ballet dancers touch iron to fend off the evil eye.

When the evil eye has struck, someone must diagnose it. In New Mexico people place a broken egg in a dish near the bed of the child victim. If the eye appears in the yolk, or the yolk preternaturally "cooks" overnight, the child is spellbound. In 1994 the American Psychiatric Association added the evil eye to its *Diagnostic and Statistical Manual of Mental Disorders* (DSM), the bible of the psychiatric community. Traditional methods had been misdiagnosing this and other culture-related ailments, like *pibloktoq*, which sends Inuit into frenzy and often brief coma, and *zar*, an African spirit possession whose victims laugh wildly, weep, and bang their heads against walls.

Once hexed, the victim needs help. Physicians are power-less. Common remedies are incantations, rites, incense, counter-spells, and special potions. Among Hindus, saliva cures most praise-induced woes. The Shilluk, a Nilotic people of the Sudan, will pay the possessor of the evil eye as much as a cow to remove the spell, yet they also fine him if he boasts of casting it. The Tamils of southern India burn ten to fifteen brooms. Malayalis call upon the bachelor-demon Vudikandan to treat the evil eye, and no one reveals his secrets to outsiders. A cat can cast the evil eye in South India, and to dispel its effects, peasants roast a fish over the head of the sufferer, then feed it to the cat.

The evil eye may be largely inadvertent, but we excel in more calculated expressions, and they too form a major part of our experience of the face.

5

Five

The Lie and the Veil

BILL Clinton speaks with more than his vocal cords. Standing before the TV cameras, he juts his chin forward, signaling determination. He pushes his lower lip out, suggesting defiance and inner timber. He narrows his eyes, showing concentration. And he smiles a great deal, conveying reassurance and confidence.

All effective presidents are masters of the face. Asked tough questions in press conferences, Ronald Reagan grinned boyishly and tilted his head, suggesting, "How important could this be?" When rebuking, his eyebrows dropped in menace. And he too smiled whenever possible. Even the dictator Benito Mussolini feigned expressions of power and aggression in his speeches, so subjects could view his fiber.

Face makes presence. In some ways, we respond to the countenances of Clinton and Reagan more deeply than their words. Their faces activate older brain circuits and sum up a mind.

Most people can't fake the basic expressions, but everyone shapes the face. Cultures also mold expression in ways individuals may not realize, and we put on faces proper to our sex and station, and to the moment. And of course people lie with the face. We rely on it so deeply that it becomes a natural playground for deceit. In fact, patterns of lying are genetic and run in families. Actors, of course, are the ultimate liars. They are lords of the facial nerve and we reward them magnificently for it.

Most such actions add signals to the face. But we also subtract them. We feign neutrality to conceal excitement or chagrin, and in many dramatic traditions, a mask covers the face of the actor. And in one nondramatic tradition, women go about masked every day, in a veil. Their faces wane away to cipherhood.

The Sultan's Silence and the Japanese Smile

Face interpretation is vital. It helps us tell a cold person from a shy one, for instance. Montaigne said, "I think there is some art to distinguishing the kindly faces from the simple, the severe from the rough, the malicious from the gloomy, the disdainful from the melancholy." We need to know who means well to us and who doesn't.

But the crux is deception. Among humans—and generally animals too—deceit comes in three basic types: opacity, camouflage, and active lying.

"Do not let the enemy see your spirit," warns Zen samurai Miyamoto Musashi (1584–1645) in The Book of the Five Rings, and the neutral face conceals it. Chinese judges wore the first smoked glasses—sunglasses—to hide their expressions during trials. Native Americans seemed stoic to nineteenth-century whites like Francis Parkman, but they were simply denying information to invaders and were fully expressive among themselves.

In a hierarchy, superiors tend to control facial information, inferiors to reveal it. The Ottoman sultans took this precept to the ultimate. When ambassadors spoke to them, they rarely responded in any way—it was like addressing an image, one emissary said—and if they did speak, their viziers quickly hushed them up.

Cool involves facial opacity. In sculpture, the Yoruba value *tutu*, coolness or composure, as did the Greeks. One anthropologist observed that among the Gola of Liberia, "The pinnacle of success . . . comes with the ability to be nonchalant at the right moment . . . to reveal no emotion in situations where excitement and sentimentality are acceptable."

Privacy is control of information about oneself. It can confer power. In poker, good players show neutral or confected faces, curtaining their inner reactions. Dictators like Stalin not only tightly manage information about themselves, with giant icons and pervasive falsehoods, but also use spies to divest others of such control. In Zamyatin's *We*, everyone lives in glass apartments, utterly visible to everyone else, except during the hour slotted for sex. It isn't paradise.

Georg Simmel suggested that some concealment is crucial in even the most intimate relationships. The attraction ends, he observed, as soon as the closeness "does not also contain, simultaneously and alternatingly, distances and intermissions." Little mysteries give intimacy its thrill.

Camouflage is one of nature's grand strategies. Ptarmigans, horned lizards, and peppered moths blend in to evade predators. Indeed, animals have mimicked almost every element of the natural environment, including twigs, leaves, seaweed, rocks, and bird droppings. Carnivores exploit the tactic too. The floating-log look of crocodiles is no happenstance and the white fur of polar bears helps them sneak up on seals.

Many animals swiftly match themselves to background color. The chameleon is the marquee creature here, but other lizards, fish, frogs, squids, shrimps, and insects also perform the trick. In fact, the tropical flounder and octopus can mimic patterns of color and texture, and will even attempt a checkerboard. The secret of these quick-change artists is often simple: They swell or shrink cells containing melanin. It's like changing dot size in a halftone.

People camouflage thoughts and emotions. We learn to accommodate social expectations of all sorts. Courtiers, like those in *Cymbeline*, "wear their faces to the bent of the king's looks," and Prufrockians "prepare a face to meet the faces that [they] meet." Culture, sex, status, age, and many other factors affect the expectations.

All cultures have display rules, what Ekman calls "norms regarding the expected management of facial appearance." These rules ordain which expression to show in a given circumstance, and how much to show it. They explain why tears flowed so freely from the Polish nobility, and why Mediterraneans seem more "emotional" than Scandinavians. Display rules can lie deep within us. Children learn some of them at the age of two or three, and one study found them in place before the first birthday.

Perhaps the outstanding example of a display rule is the Japanese smile, which Lafcadio Hearn called "an elaborate and long-cultivated etiquette." Japanese live on small, crowded islands and prize politeness as a social lubricant. Scientists find that both Japanese and Americans show the same expressions of fear, disgust, and distress when alone. With others, though, the Japanese mask negative emotions with smiles much more often. They not only smile when pleased or embarrassed, like Westerners, but also when depressed and when conveying painful or shocking information. They aim to avoid disturbing others, but Westerners can find this smile astounding and offensive. Intriguingly, the Japanese actually look less favorably on smiles overall, knowing they are so commonly false. Hence while Westerners smile in photos, Japanese more often look neutral or grave.

The Japanese also tend to suppress anger, and the Chinese could apparently faint and even die from it. In the Chinese novel *All Men Are Brothers*, a character says, "Today I am killed by anger," and succumbs soon after. Anthropologist Otto Klineberg saw a patient in a Beijing hospital whose father reportedly died of anger after losing a lawsuit. Such fatalities are fairly slow. The person turns his rage inward, falls ill, takes to bed, and expires after several days or weeks.

Some display rules follow gender lines. For instance, many cultures have tried to stamp out the belly laugh in females. Ovid, telling women how to win men, says, "Do not open your mouth too wide, like a braying she-jackass. Show your dimples and teeth, hardly more than a smile." He recommends similar restraint when feigning tears.

Japanese has terms for laughter appropriate only to women. For instance, the *hohoho* is made with a small, rounded mouth. It is vanishing, mourns one Japanese scholar, because today "Japanese women are neither graceful nor modest enough to pay attention to the way they laugh." *Ohoho* is a female-only laugh of greater restraint, often made with the hand over the mouth. It too is an antique, old men say, in the saucy modern world.

Among men, display rules often guide revelation of vulnerability. Men typically win praise for choking down signs of pain, though their eyes may still narrow when they try to hide it. In the same spirit, men cry less than women, though this rule has loosened in the last twenty years.

Occasions have display rules. Most notably, the funeral demands gloom, and at the proper level. If a dead man's secretary grieved more than his widow, for instance, she would shock mourners and kindle possibly false rumors.

Status affects facial displays. We tend to laugh less loudly in front of superiors and elders, and though we reveal more, we act more carefully. We don't want to make mis-steps. Bosses can have their own protocols. In Japan, they are chary of joyful expressions, so a coach may appear stone-faced even when his team wins a great prize.

Indeed, a person with status develops "face," which she can save or lose. The terms "saving face" and "losing face" come from East Asia, where they referred directly to dignity. In China "face" or *mien-tzu* was synonymous with honor and prestige as early as the fourth century B.C. In Japanese, *kao* means "face" as well as "honor," "status." *Kao ga tsubureru* means "to lose face," but literally it's "to crush or ruin face," and *kao ga tatsu* means "to save face," though literally it's "to establish face." The notion of "face" in this context is so rich that scholars use it rather than a more abstract term in investigating the phenomenon.

Opacity and camouflage may be the most common forms of deception, but active lying is the most interesting. And we are hardly its sole practitioners. Chimps are artful liars. In a zoo, they like to fill their mouths with water, stroll about in innocent deadpan till they near a prospect, and hose him. Even experienced handlers fall victim to this trick. When Frans de Waal began working with chimps, after years with monkeys, he admitted, "I was totally unprepared for the finesse with which these apes con each other. I saw them wipe undesirable expressions off their face, hide compromising body parts behind their hands, and act totally blind and deaf when another tested their nerves with a noisy intimidation display."

Other animals also use false signals. Vervet monkeys have a set of seven alarm calls, some quite specific, such as for the

leopard and the eagle. A monkey issues a call and the pack scat-
ters, often leaving food behind. Occasionally a monkey will fake
a call, hoping to seize the abandoned food. If he does so too
often, the pack learns to ignore him, but for that call only.

To approach a female *Photuris* firefly, a male flashes the signal of her prey. The female responds with her own counter-lie:
the enticing flash of the prey's female. As they near, the female
sees her ploy has failed, but the male has a greater chance to
mate. Genes for lying play a crucial role in propagating this
species.

Like fireflies, people have genes for lying. One remarkable
study of 1,819 Hawaiians found that, of 54 different personality
traits, family members most resembled each other in tendency
to lie. Twin studies have supported the finding and grounded it
in DNA. Family members deceive in similar ways, because of
their genes.

Some people excel at falsehood. These natural liars are
usually quite aware of their talent, since they have deluded parents and teachers to escape punishment since early youth. They
are confident and feel no fear or guilt about getting caught. Yet
they are not sociopaths; they don't use their skills to hurt other
people. In fact they score the same as other people on psychological profiles. But they would seem to do better in certain
careers, like sales, diplomacy, politics, acting, and negotiating.

Other individuals, true Pinocchios, seem incapable of hiding a lie. Their very worry about detection exposes them, and
makes them fret about lying in the future. They and most of
their friends usually know of this trait, and it can be a true social
handicap. Again, these people score no differently from others
on personality tests.

But with most people, detecting a lie requires art—or a
dash of science.

How to Spot a Lie

"I think there's never a man in Christendom can lesser hide his love or hate than he," declares Hastings of Richard III, "for by his face straight shall you know his heart." It is one of the great naive utterances in literature, for moments later Richard orders his death.

Deceit obsessed Shakespeare. It propels *Othello* and *Lear*, pervades *Richard III*, undergirds *Hamlet*, and dances throughout his romantic comedies. Had the Bard's characters known how to identify lies, these plots would deflate, just as rhinoplasty would kill *Cyrano de Bergerac*.

But we are surprisingly poor at it. As Walter Mosley's detective Easy Rawlins observes, "Some people say that they can tell when a man is lying. Those people are fools. You can never trust what someone tells you is true." Study after study has found that people are 45 to 60 percent accurate in spotting lies — close to chance's 50 percent. In one typical study, only 3 of 109 observers scored over 70 percent.

Even so, we preen ourselves on our acuity with lies. In one experiment, police detectives performed no better than ordinary citizens, but to the researchers' surprise, they were confident they had. Another, in which half the subjects were lying, found the following mean correct guesses, by percentage:

Secret Service agents	64.12
Psychiatrists	57.61
Judges	56.73
Robbery investigators	55.79
Federal polygraphers	55.67
College students	52.82

Only the Secret Service agents scored significantly above chance, possibly, investigators speculated, because they were used to scanning faces in crowds. In all groups, the subjects' self-assessment of skill at lie detection bore no relation to actual score.

A recent Dutch study asked prisoners, police detectives, patrol police officers, prison guards, customs officers, and college students to name the factors that indicate lying. The prisoners proved significantly more astute than any other group, perhaps, researchers said, because they live in a culture of deception, full of posing, bluffs, and cons, and hence get more feedback.

Probing for the truth can actually blind us to lies. In one study, neither suspicious nor accepting questions boosted accuracy, and in another, probing lowered scores. When pressed, liars increased eye contact and facial animation, convincing probers of their honesty.

Across cultures, display rules deepen the opacity Americans and Jordanians proved completely unable to recognize each other's falsehoods in one study. Among Japanese and American trade negotiators, the Japanese make much less eye contact and are often silent for longer periods. We associate gaze aversion and hesitation with deception, a fact one researcher suggests may worsen trade problems.

Moreover, we pay most attention to the worst cues and ignore the best. For instance, we value verbal content too highly. It can yield solid proof of mendacity, but liars mold it easily. We think the jitters signals deceit. It can, but some people fidget naturally and others from fear of disbelief, even when speaking the truth, while still others utter lies from an aura of confidence. We deem an averted gaze a sign of dishonesty. In fact, we do tend to look down and away out of guilt or shame — as we look straight down with sadness, and to the side with disgust. But eye gaze is a perfidious clue. Liars anticipate both nervousness and averted gaze, and mask them. At the bigamy trial of Giovanni Vigliotto, who may have had a hundred wives, one of them testified she was attracted to his "honest trait" of gazing straight into her eye.

Leibniz suggested study of the face and voice could enhance lie detection. Paul Ekman, in his book *Telling Lies*, notes that a few people have achieved an accuracy rate of 80 to 90 percent, and they do in fact rely heavily on facial and vocal signals.

Which ones? In fact, no single cue reliably flags a lie. "Rules of thumb don't work," Ekman explains. "Anyone who can say, 'I know what to look for,' is going to be wrong most of the time." But rise in pitch is one excellent tipoff. Lying voices climb the scale, for reasons not entirely clear. Prevaricators also touch themselves more, much as chimps groom themselves more under stress. Other cues include: longer time before answering, briefer answers, more *um*'s and *uh*'s, slower speech, and fewer hand and arm gestures to illustrate points.

In the face, the situation is more complex and fascinating. We are all "two-faced." We come with one neural system for involuntary expressions and another for deliberate ones. Hence some patients can have lesions which kill all their voluntary expressions, but not involuntary ones. Their faces are truly transparent, their expressions candid. Others have the reverse, and can show only willed looks. Both systems seem to develop separately. Their interaction is complex, and it provides cues to lying.

We have much greater control over the muscles of the lower face than the upper, probably because we need them for eating and speech. So the lower face is more adept at deception.

A key clue is the smile. The Ananias shows fewer felt smiles and more masking or cover-up smiles, and studies show that these, along with higher pitch, are the best evidence of deception.

A false smile stems from lack of emotion. In baldest form, it's scarcely even deceit. The eyes seem vacant and the lip corners lift in a wan simulacrum of pleasure: Here is what I'm *not* feeling.

Yet a false smile can also be harder to detect, and six giveaways reveal it.

First, it uses just the *zygomatic major* muscle and hence animates only the mouth. The *orbicularis oculi*, the cheek puller, lies still, so the cheeks are slack and the eyes don't narrow. Artful liars compensate by contracting the *zygomatic major* intensely. That act alone affects the *orbicularis oculi*, swelling the cheeks, narrowing the eyes, making the false smile look more credible.

False smiles can endure too long. True smiles last between two-thirds of a second and four seconds, depending on intensity. But false smiles often linger on, like awkward guests after a party. Since these smiles lack an authentic inner prompt, we don't know when to retire them. In fact, any expression over ten seconds long, and most over five, are probably fake. Intense displays like rage, ecstasy, or depression are exceptions, but even they usually appear as a series of briefer displays.

While felt smiles fade away naturally, false ones can seem guillotined. In fact, an abrupt start or end to most expressions shows we've issued a conscious command. Surprise is the exception, a flash. It always has onset, duration, and offset of less than a second, and if it persists longer, the person is counterfeiting. Many people can mimic the look of surprise — the raised eyebrows and open mouth — but few capture its rapid start and end.

Fourth, false smiles tend to a slight asymmetry. The left lip corner rises higher on a right-hander's face and vice versa. Indeed, almost all deliberate looks share this quality. In false anger, for instance, the left brow lowers more, and in feigned disgust, the nose wrinkles and the lips pull back more on the left.

This lopsidedness occurs because the cerebral hemispheres control willed expressions, and the dominant one sends a stronger signal. Involuntary expressions, in contrast, arise in the lower, unitary brain and affect both halves equally. Asymmetry is very subtle, but people can often spot it in the lab if asked to look for it. Like all other cues, however, it is not a master key. A few deliberate looks are symmetric, and a few natural ones are not.

The smile that arrives too early or late is probably an imposter. The inner cues are absent, so dissemblers can botch the timing. For instance, if a person says, "I've had it with you!" and then shows wrath, it is likely a confection. Likewise, facial expressions should occur with body gestures, not after them. The person who bangs the table then looks angry is playacting.

Finally, masking smiles cover up grimmer emotions. When Count Dracula says he only wants solitude in London, he smiles and Jonathan Harker senses masking: "Somehow his words and his look did not seem to accord, or else it was that his cast of face made his smile look malignant and saturnine."

The hidden feeling often leaks in the upper face. "The eye is traitor of the heart," said poet Thomas Wyatt (1503–1543), and the eyes, eyebrows, and forehead abound with reliable muscles. Here lies the true Pinocchio's nose in people. A person who tries to conceal grief behind a smile will often give herself away at the eyes, where sorrow speaks. As one recent study found, when cues from the mouth and eyes conflict, the eyes prevail.

The forehead is the capital of these muscles. For instance, with sadness, distress, grief, and guilt, the inner eyebrows rise and wrinkle the center of the forehead. Fewer than 15 percent of people can fake this gesture. Even harder to manufacture is the telltale sign of fear or worry: the rise and pulling together of the eyebrows. Fewer than 10 percent of people can mint it.

The involuntary muscles are good tipoffs elsewhere too. For instance, grief and sorrow pull the lip corners down without moving the chin muscle. Yet fewer than 10 percent of people can perform this act at will. And tightening the lips is a dependable sign of anger. It often heralds the emotion and is, essentially, a warning of a warning. The lips don't pull in or press together markedly, but rather slip into a subtle sternness. People can learn to mimic such expressions, but it usually takes hundreds of hours.

To cover fear or guilt, shrewd liars often use another strong emotion, like indignation. In 1960, Israeli interrogators of Adolf Eichmann soon learned that whenever he burst out, "Never! Never! Never, Herr Hauptmann!" or "At no time! At no time!" he was lying. At the Thomas-Hill hearings, Anita Hill spoke softly and articulately, while Thomas fumed. His indignation may have covered lies, though alone it was hardly diagnostic.

One secret of facial truth, known to almost no one, is the micro expression. For an instant or two, the face completely flashes the emotion it's hiding. Tape analysis of the Iran-Contra hearings shows that John Poindexter, questioned about a meeting with CIA head William Casey, displayed a picture-book face of anger for a quarter of a second. It meant the apparently unflappable Poindexter may have been hiding a key fact about this meeting, but interrogators missed it and didn't follow up.

Untrained people normally don't notice micros, though they spot them on slow-motion videotape. But clinicians detect them readily, and most individuals can learn to recognize them with an hour of training. Indeed, people can teach themselves, just by flashing photos of facial expressions before their eyes and trying to guess them.

The gemlike micro is fairly rare. A messier phenomenon is much more common: the squelch. A damning expression starts to emerge, the person senses it, and covers it rapidly, usually with a smile. A squelch can occur quickly enough to hide the underlying emotion, but it usually lasts longer than the micro, and even when we can't see the emotion, we often sense the squelch itself.

What about rapid blink rate and dilated pupils? Liars often exhibit them, but both are general responses to arousal. Besides prevarication, they can signal excitement, anxiety, fear, anger, or other high emotion. Yet blinks especially can be useful, and they work best as cues when strong feeling is out of place for the speaker.

With all these giveaways, we remain slow to detect lies. If we're so bad at spotting them, why don't they proliferate? Minor lies have, to our advantage. For instance, they can be social grout. The false smile fosters friendship, softens blows ("I appreciate the joke, even if it wasn't funny"), smooths social bumps and potholes. The lying pleasantry—"I had a great time" from the escaping party guest—can benefit both deceiver and deceived.

Flattery is a lie by definition, and fools us with splendid ease. Praise resonates with our self-image, seems intuitively true, or at least gives no incentive to disbelief, and underlings have always found clever sport in exaggerating it. Rameau's nephew, a professional toady, brags of his art to a disgusted but rapt Diderot. He claims "an infinite variety of approving faces, with nose, mouth, eyes and brow all brought into play," and Diderot calls him "one of the weirdest characters in this land of ours where God has not been sparing of them." But receipt of flattery establishes our dominance and profits us—up to the point where, as in *Lear* and many Roman comedies, it hides treachery.

Other kinds of useful lie abound. A placebo is a lie, and psychologists think it works by reducing stress. An epitaph is often a lie. Deception researchers like Ekman lie to subjects, misleading them about their privacy or the aim of the experiment, and indeed lies sustain much social science research.

Most lab studies have dealt with minor lies. But as the stakes rise, prevaricators stumble more: the "motivational impairment effect." This effect peaks in people with low confidence, like interviewees who desperately want a job and lie to try to get it. These individuals apparently try to control the total message — face, words, vocal tones, gestures. It's like thinking about how to walk, and about as helpful. Such individuals seem awkward and insincere even when telling the truth. Men bumble in this way more toward women, and vice versa. Face and hand gestures especially betray these falsehoods; we identify these liars better on a muted videotape than an audible one. The motivational impairment effect makes evolutionary sense, since important lies cause more damage. If everyone could utter them fluently, the ability to detect them would become a genetic prize and spread widely, as it apparently has.

And, finally, we give people the benefit of the doubt, since they usually do speak honestly. Indeed, our species evolved in bands of perhaps 30 people each, loosely united into wider tribes. In such small groups, lying would be much riskier, because the exposed liar would have to live with her stigma. Moreover, people could more easily discover lies through facts known to others. Richard Wrangham, who studied African pygmies, found that a husband could best prevent his wife's infidelities by making it clear he knew the gossip of her friends.

Even today, people guard against more harmful lies by basic fact-checking. As if acknowledging our weakness before liars, we contact personal references, verify credit card limits, hold banknotes. And we punish the deceitful. We lose trust in people we catch in just a few lies, and we can cease dealing with them. We also spread the word about liars, who may then find themselves in a chilly sea of suspicion. Stores circulate names and photos of bad-check artists, and computers keep meticulous track of people's misdeeds. The ultimate penalty for lying is ostracism, which does not spread one's genes.

Lab studies in this area have limits, as researchers themselves point out. For instance, we spot deceit better in those we know, yet most studies involve strangers. Indeed, after subjects view even one of a person's honest utterances, they discern his lies more adroitly. In everyday life, we may penetrate falsehood better than the studies suggest.

Yet we are hardly proficient, and lies have always been a chariot to power. Machiavelli advised the prince to be "a great feigner and dissembler," and in 1908 Simmel called the lie a way of controlling the less insightful and suggested that it operated "as a selecting factor to breed intelligence." Persuasion is often deceptive or semideceptive as well, and its adepts win power, wealth, and status. As Swift said, "If a man had the art of the second sight for seeing lies, as they have in Scotland for seeing spirits, how admirably he might entertain himself in this town." His London could be any town.

Universal Faces

Thomas Gainsborough found the faces of David Garrick and his rival Samuel Foote too phantasmagoric to paint. No expression lingered long or summed them up. "Rot them for a couple of rogues," he groused. "They have everybody's face but their own."

But that is the art of actors. They tame the involuntary expressions and shape the face on cue. Janus had two faces but an actor has hundreds. Indeed, a mute performance like Holly Hunter's in *The Piano* is a triumph of face.

The skill seems preternatural, and we routinely elevate actors to a semi-pantheon. Romans called any master of his profession a "Roscius," after the prime actor of the day. Despite Gainsborough, painters depicted David Garrick more often than any of his contemporaries. and modern actor-celebrities keep the paparazzi clicking and the tabloid presses awhirr. It's understandable, for no art is more immediate, and actors use themselves as a medium in a way sculptors or musicians do not.

Yet there is also a wonder to acting.

The nub, of course, is mimesis. How can a person summon rage, tears, and laughter at will? ("And all for nothing!" cries a disgusted Hamlet. "For Hecuba!") Beneath that riddle lie others, which Denis Diderot broached in *The Paradox of the Actor* (1773). How do actors keep a thousandth performance green? How do they achieve "prepared spontaneity," the oxymoron at the center of the art? How do they perform believably regardless of mood? How do they simulate grief or horror without feeling it?

Or *do* they feel it? Some early theorists saw the actor as a fount of emotion, overflowing into the part. Horace said, "We laugh and weep as we see others do,/ He only makes me sad, who shows the way/ And first is sad himself." Or, as playwright Robert Lloyd (1733–1764) stated, "No actor pleases who is not first possess'd." An actor pours himself into Macbeth or Mosca, filling out a face with his heart.

But Diderot said actors onstage felt nothing. For that very reason, they could counterfeit all feelings. David Garrick could run through an arpeggio of emotions in five or six seconds because he stood apart from them. True feeling was an obstacle. It overwhelms and somewhat incapacitates us, he noted, but actors adapt to the moment. Like all artists, they stay in control.

Diderot pointed out other difficulties with the cup-runneth-over theory. A person acting from passion would perform less credibly over time as the feeling waned, but actors improve. After a tragedy the audience is sad, but the actor is simply weary. An emotion-bound actor would flounder in times of personal distress, but actors go on. In love scenes, actors sometimes bicker *sotto voce* in between their amorous sighs and avowals, eerie if real passion were moving them.

Of course some of Diderot's critique reflected the rule-laden nature of the French theater, with its singsong declamations and rigid facial conventions. The latter stemmed from *A Method to Learn to Draw the Passions* (1702) by Charles Le Brun (1619–1690), court painter to Louis XIV. Le Brun adopted Descartes's scheme of six fundamental passions—love and hatred, joy and sadness, admiration and contempt—and described their expressions. The recipe for contempt, for instance, was: eyebrows knit and slanted inward, eyes wide open with the pupil in the center, nostrils flared, mouth shut and its corners drawn down slightly, lower lip thrust out.

Likewise, in *The Art of Acting* (1746) dramatist Aaron Hill (1685–1750) laid out ten basic passions the face could express—joy, wonder, love, pity, jealousy, fear, grief, anger, scorn, and hatred—and limned each one. An actor shouldn't utter a syllable until he had mastered them. Anger had a single look rather than many, and actors bragged of their ability to summon it.

These Platonic faces proved suffocating. In England, Charles Macklin and David Garrick deployed the cornucopia of expression, fascinated audiences, and shattered this little game. They brought stage characters to life.

Yet Diderot's questions linger on.

Do actors express emotions without feeling them? Simon Callow describes the sensation from an actor's viewpoint: "No matter how intense or painful the emotions of a part, the more you enter into them in a good performance, the less you are affected by them. It is they and nothing else which are activated. The emotion passes through you." The question may not be whether actors feel, but whether the feeling digs into the soul. Normal emotions do; these glide on by. The anguish of Blanche Dubois or Willy Loman is not the anguish of the actor.

If not, what is going on? How do actors fool us with showcase emotions?

One secret is thinking. "Just rely on your character's thought processes and your face will behave normally," advises Michael Caine. An actor who thinks the character's thoughts will seem to mint lines freshly. He'll be acting inside and out, and the process itself will create the right face — as well as the voice, pauses, gestures, and movements of the character.

Mastering this thoughtstream is a painstaking business. Acting requires careful script analysis and many rehearsals. Initially an actor simply reads lines and gives a shallow performance. It's impossible to read a part and think it at the same time. Once he memorizes it, though, he frees mental acreage for acting and performance soars.

But memorization is not enough. The lines must sink into the cortex like the Pledge of Allegiance, so they spool out by themselves. Paul Muni practiced until the lines came reflexively, then junked them and simply thought the ideas they conveyed, often straying from the script. Then he let the lines return. In real life, few of us arrange words in our heads, then utter them. We speak thoughts and the words flow naturally. Actors seek the same effect.

There's more to this craft, of course. Thinking alone doesn't explain how actors don whole new personalities or summon violent rage and grief. One way of evoking wild passions emerged from the first formalized teaching of actors.

★   ★   ★

In the summer of 1906, the famed Russian actor and director Konstantin Stanislavski (1863–1938) was in Finland. One day he went out to a cliff overlooking the Baltic, sat down, and let his mind rove over his career. Certain moments in parts had defied him, and he wondered why. He worried he was growing slothful, sinking into rote repetition of roles. These doubts led to years of self-examination and ultimately to the instruction technique called "system."

Stanislavski wrote copiously and multifariously, and in the end his system eludes systemization. But three points stand out: relaxation, concentration, and emotional memory.

John Gielgud called relaxation "the secret of all good acting." Actors normally feel tension and it can petrify performance, cutting off facial reaction to thoughts. Relaxation quells it. A calm actor becomes physically supple, and her expressions flow freely. Marilyn Monroe was wooden before the camera until she won her first fame. Then she learned to relax before it — indeed, to snuggle up to it — and her screen presence bloomed. In life and art, relaxation bestows grace.

Concentration is also crucial. As Stanislavski observed, an actor onstage is "private in public," an attitude that demands focus. An actor plainly aware of the audience forfeits its attention, yet one who is quiet and still, but intensely concentrating, can hold an audience rapt. Actors train their focus by such exercises as working imaginary clay or richly envisioning the first bed they remember, the room, the sounds, the air temperature, the blanket pattern, even the dust particles in the slanted light.

Emotional memory is the most famed and arresting facet of the system. In 1958 physician Wilder Penfield found that touching electrodes to the brains of patients in surgery brought back vivid, you-are-there recollections. One woman, for instance, saw herself back in her kitchen, worrying about her child outside near a busy street. She remembered not just the room, but how she felt at the time. Past experience is a mix of event and feeling, and our minds recall both.

This interesting fact means voluntary memory can unleash involuntary emotions. It is a key to the control room.

The Stanislavski system exploits it. By recalling the death of a parent, for instance, an actor can make tremulous grief well up in the face. James Earl Jones described the technique as building an emotional track parallel to the character's. He could not directly recreate the jealous death-spiral of Othello, for instance, never having experienced it. But he could build an imaginary track based on his own memories and use that.

It works, as Paul Ekman confirmed for science. He asked subjects to make involuntary expressions, and when they failed he suggested they call up an emotional memory. They then often succeeded. But Stanislavski's students knew it along ago, since their classroom exercises in emotional memory could leave them weeping, snarling, or laughing uproariously.

Lee Strasberg (1901–1982) refined the system and in his hands it became the Method, which he taught to such actors as Robert DeNiro, Dustin Hoffman, Paul Newman, Joanne Woodward, Al Pacino, Marilyn Monroe, and Marlon Brando. Strasberg believed that, though an actor had to think as she delivered lines, it often didn't matter what she thought about. She might be contemplating whether to eat at Sardi's or Pizza Hut. Her face would seem alive anyway, partly because her brain was working freely.

But the Method is not to every actor's taste, and many treat it as just another tool in the kit. Laurence Olivier felt it worth knowing, but also believed it delayed rehearsals and buried the natural gifts of Marilyn Monroe. Jones himself said he found it useful, but no more. There is no Euclid for acting.

Acting changes from stage to screen, where the face grows even more important. As George Cukor told Katharine Hepburn, the camera sits right by the actor's shoulder. "The camera doesn't have to be wooed; the camera already loves you deeply," says Michael Caine. "Like an attentive mistress, the camera hangs on your every word, your every look; she can't take her eyes off you."

Indeed, it is clairvoyant. It scans the brain, catching tiny flickers of effort and hesitation and magnifying the blankness of an uninhabited role. This all-seeing power grows especially keen in close-ups, when an actor's eyes show his mind and he must think the character's thoughts with utmost concentration.

Adroit reaction is another requisite of screen acting. Spencer Tracy admonished young actors, "Take your hands out of your pockets and listen to the other actors." Audiences sense good listening subliminally, and it brings conversation alive. A skilled actor scrutinizes another actor's lines in advance to pinpoint when they trigger her thoughts. A single word halfway through may spark her next line, though she can't speak it until the other person finishes. Yet she can think it, and the camera will see her mind at work in her face.

A clever actor will go even further. Like a real person, she can have excellent things to say that never quite reach her lips. It doesn't matter. The camera will see them, and so will the audience. The thinking face of a performer like Ben Kingsley, for instance, is rich and mesmerizing. Emma Thompson excels at this device, and it gives many of her characters a whimsical inner world. "To *act* thinking is hard, impossible, in fact," says Callow. "You must *think*—that's all there is to it."

On stage or screen, it all takes time. "A film actor must be able to dream another person's dreams before he can call that character his own," says Caine. John Gielgud said he needed six weeks onstage to gain full command of a part.

Ekman observes that people can learn involuntary expressions, though it takes hundreds of hours. Actors log in far more than that. Gielgud felt it took a minimum of fifteen years to make a good actor, and for most people it demands a lifetime of practice.

It may also require a genetic gift and perhaps special topologies of character. Diderot advanced a chameleon theory here: Actors had no character and thus could play any character. Director Peter Brook said most actors wished to write off their basic natures by the age of ten and used acting to dwell in fantasy personalities they preferred to their own. Leo McKern said acting is hiding and Gielgud wrote, "For me the theatre has always been an escape, a make-believe world, full of color and excitement."

Yet actors also seem to teem with inner selves. Callow and Dame Edith Evans have said that many personalities jostle about within them. Such an inner crew can obviously help a performer take on an array of personalities, though serious actors go further and examine strangers in the street, study other actors, and become participant observers in special settings.

How do actors do it? There is a story about Laurence Olivier. He was playing Othello and had the audience spellbound. At the curtain call even the cast applauded him, but he

stalked to his dressing room and slammed the door angrily
behind. An actor crept up, rapped timidly on the door, and
asked, "What's the matter? You were fantastic!"

"I know!" he shouted. "And I don't have one goddamned
idea why!"

Proteus Bound

In Victor Hugo's *The Laughing Man* (1869), Gwynplaine seems
locked in mid-guffaw. "His face laughed; his thoughts did not,"
says the author, who speculates that he'd undergone an operation in infancy to make him a lucrative sideshow exhibit. "One
might almost have said," Hugo notes, "that Gwynplaine was
that dark, dead mask of ancient comedy."

A real frozen face like Gwynplaine's excites a shudder, yet
intriguingly one in drama does not. A one-face actor might
seem no actor at all, but playwrights have used a passel of tricks
to make expression well up from the mask. Even so, masks force
drama into special shapes.

Masks have been surprisingly prevalent in drama and may
in fact have spawned it, via ritual. Africans commonly wore
masks in rituals, and some of them came alive only when used,
a bit like a ventriloquist's dummy. In Chinua Achebe's novel
The Arrow of God (1964), the carver Edogo worries about his
most recent mask. Clients praise it, but he knows he must see it
in action to know its true worth. When the ceremony begins,
"the weakness seemed to disappear," and the mask springs to
life. An evening fire often aided this Hoffmannesque process,
since it made the shape seem to fluctuate.

From donning a spirit's identity, it's a short step to acting
out its deeds. Masks de-face drama, but they've appeared in plays
around the world. Such plays abounded throughout Southeast

Asia, and among the Pueblo and Hopi Indians. In Aztec plays, actors wore masks of mythic figures as well as of toads, birds, beetles, and butterflies.

In China, the mythological romance *Fung shen yen i* required twenty-one masks. Some depicted deities like Yün Siao, whose powerful scissors cut men and gods in half, or the mythic Kiang Tse-ya, who was so compassionate he fished with a straightened hook. In Chinese Buddhist dramas, viewers saw the agonies of hell. The Chinese inferno has myriad hells in ten levels, through which the soul travels on its way to reincarnation. There are steamy, icy, and gloomy hells, hells whose denizens drink ghastly drugs, or sit on pikes, or slip endlessly on a path of oily beans. Masks helped render these tortures.

But the Japanese and Greeks took this art to its acme.

Noh drama arose in the 1300s, developed mainly by Zeami (Yusaki Motokiyu, 1363–1444) and his father from earlier forms. In 1374 the shogun-esthete Ashikaga Yoshimatsu attended a Noh play. He liked it so much he blessed it with his patronage, and from then on it flourished. Later, the Tokugawa shogunate turned Noh into a state-subsidized art for the elite. It typically treats the deeds of feudal Japan, and is formalist, anti-realist, and often hauntingly elegant.

"A mask tells us more than a face," said Oscar Wilde perversely, and some Noh actors might agree. Over 500 masks exist for the parts in Noh. They come in five types — men, women, gods, demons, and madmen — and an array of subtypes. Like Africans, Noh actors seek to make these small, lightweight masks live, and when an actor successfully animates them, they seem to melt into the face.

Their subtle contours aid the feat. The mask aims at inner essence, Zeami said, and should "stylize the character it represents and show it in a truer light than reality could ever do." The best catch a look that quivers on the brink of several possible expressions. They are so evocative that, before performances, many actors stare deeply into the mask to imbibe its spirit.

Artists carved Noh masks with a keen eye to lighting, and it is a second secret of their power. A lady pining for her distant swain lowers her head, and the angle of light deepens the shadows in the mask, makes it seem more sorrowful. A warrior freed from danger lifts his head to the light and the mask exults. Sometimes a simple turn of the head can curl the lips into cruelty.

Yet they are still masks, and lack the full vocabulary of faces. Noh compensates with a code of emblems. Since a mask can't weep, the actor shows tears by moving the right hand slightly up and down over one eye, then the other. To express strong joy or anger, he stomps his foot to the beat of drums. As in ancient Greece, a chorus further clarifies characters' feelings for the audience.

Noh is elaborately stylized. In addition to masks, the actors wear awkward, multilayered costumes, and walk and gesture according to strict formulas. Masks may force stylization on all drama, but in Noh it reaches a sublime dead-end. One must progress by retreating.

Aristotle says Greek tragedy arose little by little from "dithyrambs," a term we do not fully understand, but which apparently involved religious rites. Greek masks were very different from the delicate Noh confections. They were large cylinders designed to let spectators see the face from the back of the amphitheater, three hundred feet away, and had metal mouthpieces to make the voice carry. The Noh mask served refinement, the Greek democracy.

The Greeks didn't try to make their masks live. Instead, they tried to circumvent them. Their plays abound in cries of lamentation, a substitute for tears. Characters specifically mention flush, pallor, and hair standing on end, so playgoers could imagine them. Actors contributed with bold gestures. In grief, they tore their hair, scratched their faces with their fingernails, beat their bodies, ripped their clothes, and, intriguingly, veiled their masks. In disgust they spat, and in joy they danced and leapt like springboks.

As in Noh, the masks of Greek drama bestowed certain boons. They could reveal character. Their stasis could be oddly bewitching, and speech from a motionless mouth added a touch of the preternatural. Masks cut the payroll, since one actor could handle several roles. For instance, Aeschylus' *Agamemnon* has six speaking parts, but only two actors played them. And masks kept actors anonymous and tractable, though in later Greek drama some actors won public recognition with their distinctive voices and gestures.

The Greek tradition endured for many centuries. Medieval mystery plays required masks, as did *commedia dell'arte*—no problem for a society that made even its portraits into masks. But as verisimilitude rose in popular esteem, cosmetics came to replace masks. By the time of Shakespeare, they were gone.

Occasionally they re-surface in the West, as in Eugene O'Neill's *The Great God Brown*. O'Neill had seen photos of African and Asian masks, and, dizzied by the ideas of Nietzsche, had swallowed whole one of his comments: "Every profound spirit needs a mask; nay, more, around every profound spirit there constantly grows a mask, owing to the constantly false, that is to say, superficial interpretation of every word he utters, every step he takes, every sign of life he manifests." O'Neill saw that masks could express this split between the complex inner life and the emoticon personality seen by the world. He also realized they could show misunderstandings, evolution of character, false fronts, even transfer of personality, as when the flamboyant Dion dies and the maskless Brown steals his mask. O'Neill himself had absorbed the personality of his older brother Jamie, an act which gave the shy young man the carapace of a rebellious skeptic and drinker, and rifted his soul. He later said, "One's outer life passes in a solitude haunted by the masks of others; one's inner life passes in a solitude hounded by the masks of oneself."

When *The Great God Brown* debuted on January 23, 1926, it had rushed into production and the masks were malformed and gawky. O'Neill felt they conveyed only hypocrisy and defensive personas, and was enraged. The spectacle bewildered the opening night crowd, but the play ran for eight months. O'Neill later recommended masks for *Hamlet*, and said other plays of his, like *The Emperor Jones*, would profit from them. "At its best," he said, "[the mask] is more subtly, imaginatively, suggestively dramatic than any actor's face can ever be." It was a quixotic opinion in his own day, and in an age of film it feels incomprehensible.

Yet film has its own masks. They are not a stylizing challenge, but a mimetic aid. Lon Chaney played the Wolf Man in yak fur and a rubber snout, but even in the 1930s makeup artists were devising the foam latex masks called "appliances" for movies like *The Wizard of Oz*. Appliances change appearance, but hug the face and preserve expression. The makeup person takes a life-mask of the actor and molds the appliance to it, so it fits his countenance. Its edges merge smoothly with the actor's face or neck, and cosmetics erase any visible seam.

Since skin bags over time, foam latex can send a character deep into his future. It aged F. Murray Abraham as Salieri in *Amadeus* and turned Dustin Hoffman into 111-year-old Jack Crabb in *Little Big Man*. Marlon Brando's appliance in *The Godfather* gave his jowls their impressive droop, and the technology put decades on Max von Sydow as Father Merrin in *The Exorcist*.

Some of these masks confer a new identity. They made Robert de Niro a semi-android in *Frankenstein* and Jeff Goldblum an arthropod in *The Fly*. These creations can be elaborate. Goldblum's required him to sit still for ten hours every day. They all achieve the Noh goal of invisibility without the actor's effort. They hide the face, but not the mind.

But there are places where people must hide their faces and their minds, every day in the street.

Fear of Eyeing

Women are dying in Algeria because of their faces. Religious zealots murdered two women at a bus stop because they weren't wearing the veil. Fifteen Muslim militants kidnapped and raped a mother and two daughters, then slit their throats — the notorious "Kabyle smile" — because they attended a school where females could go unveiled. Ironically, these women were devout and kept purdah.

Purdah is the world's harshest dress code, literally a "non-display rule." Its severity in many Muslim countries can shock a Westerner. In revolutionary Iran, police arrested women who let strands of hair slip into view, and some who defied the code, even foreigners, earned 80 lashes. In Saudi Arabia, self-appointed purdah police once abused a Saudi princess when they discovered her walking with a maid who lacked a face veil. Veiled women in the Mideast can see the sun so rarely they get rickets.

Muslims call the prescribed women's dress *hijab*, literally, "curtain." Its roots run deep. When Reza Shah banned the tent-like *chador* in 1935, in his attempt to modernize Iran, many pious older women refused to go outside, and some Iranian men forbade their daughters to attend school. Indeed, the deeply religious equate facial exposure with nudity. "To feel decent and honorable, a Western woman covers much of her body when she goes out in public. The Sohari woman covers body and face," writes Unni Wikan, a woman. "A Western woman need not feel shy to show herself naked before her husband; a Sohari woman need not feel shy to uncover her face in his presence."

A loose synonym for purdah is "the veil," though only the devout or denizens of purist states like Saudi Arabia wear a true veil. Generally, females cover the whole body except the face and hands, and sometimes they leave only the eyes exposed.

But purdah varies widely among Muslim nations. It is rare in the ex-Soviet Muslim republics of Central Asia. In Malaysia, many women dress as they please, and even wear curve-hugging sarongs that mock the puritanism of *hijab*, though they say it is enough that the garment touch the ankles.

The Tuaregs, the sly desert raiders of southern Algeria, actually reverse the pattern. Women go bare-faced and men wear the *chèche*, a turban wrapped around both head and face, exposing only the eyes. Tuaregs can recognize each other by the pattern of folds in their *chèches*. They also wear blue robes, whose dye once bled so badly it colored their skin blue. The Tuaregs used this gear to ward off sun and dust as they picked off caravan stragglers, but Tuareg men also insist they gain power by cloaking their facial expressions. Women, they scoff, have nothing to hide. Other Muslims look askance upon these people, and in Arabic "Tuareg" literally means "the abandoned of God."

As the extent of purdah varies, so does the attire. There are two main kinds of garments: body tents and face veils.

The best-known body tent is the *chador*, which de Maupassant said looked "like death out for a walk." It is a black swath of fabric which hangs from pate to anklebone. Women usually wear it over pants, a calf-length shirt, and hood, so the whole outfit is stifling. It appears mainly in Iran and among

Lebanese Shiites. The *abayya* is the Arabian *chador*, a black cloak with arm slits. The *chadris* is the Afghan version, a colorful tent with an oblong lattice over the face, and the *farshiyah* is the white coverall of Libya that leaves a single eye exposed.

The common face veil is call the *niqab*, and it completely occludes the face. The *burqa* is a more artful variant, a black-and-gold face mask of the Persian Gulf that usually hides all but the eyes. It is made of leather, canvas, or stiff fabric.

Oddly enough, given the extreme vigilance, the Koran doesn't prescribe purdah. It mentions *hijab* in this context: "If you ask [the prophet's] wives for anything, speak to them from behind a curtain ["*hijab*"]. This is purer for your hearts and their hearts." Why should believers speak to Muhammad's wives from behind a curtain?

The prophet had this revelation soon after he married the beautiful Zeinab, his most controversial wife. She had previously wed his adopted son, and one day Muhammad glimpsed her half-clad body. He slipped quickly away, but the adoptee surmised that Muhammad desired Zeinab and divorced her. Muhammad then married her, to general scandal. Soon he had another revelation: adoptions were invalid. Hence he had not actually married a son's wife. At the same time, he hid his wives behind a curtain, where he felt they would be safer.

In any case, this part of the Koran refers only to Muhammad's wives. Regarding other women, the book says, "Tell the believing women to lower their gaze and be modest, and to display of their adornment only that which is apparent, and to draw their veils over their bosoms."

These words hardly command a draconian dress code. But the ambiguity creeps in with the phrase "to display of their adornment only that which is apparent." Muslims hold that this curious line refers to a woman's face and hands, so only these should be visible. They must hide the rest of their body, except from other women, closely related men, young boys, and "male attendants who lack vigor." Very devout women go further, and glove their hands and veil their faces, an act that seems to run counter to scripture, which does suggest some display.

Beyond the dictates of the Koran, advocates of purdah offer a spectrum of rationales. Many treat it as a bouquet to women. It frees them from the leers of men and the whims of fashion, they argue, and by blanking out their looks it highlights their character, like cyberchat. The veil is also, say some, an assertion of cultural pride, a symbol of resistance to the West.

But beneath it all lies sex, like a great damp animal. According to Ibrahim Amini, an Iranian cleric, the *chador* means wives can "rest assured that their husbands, when not at home, would not encounter a lewd woman who might draw his attention away." Indeed, the threatening allure of women seems ubiquitous, and the deeper one's devotion, apparently, the more dangerous the temptation. Some ayatollahs deem the female voice so exciting they keep women silent in mixed gatherings. Geraldine Brooks's *Nine Parts of Desire* (1995) takes its title from a remark by Ali ibn Abu Taleb, founder of the Shiite sect: "Almighty God created sexual desire in ten parts; then he gave nine parts to women and one to men."

Ultimately these arguments are beside the point. Dress codes stem from custom, not sense, as fashion designers well know. And purdah fits into a larger system of sexual segregation.

This segregation varies in its strictness. In cities like Cairo, women commonly hold modern office jobs. Even in the Iran of the ayatollahs, Khomeini ordained that females could work, since the Koran held only that the prophet's wives should stay at home. But women had to wear *hijab*, and the sexes couldn't mingle. As a result, demand soared for female hairdressers, doctors, reporters (to cover women-only events like sports), professors. Women even made inroads in the Iranian parliament, which does hold elections, though all candidates need a nod from the religious elite to run.

In Afghanistan, the Taliban pulled the knot tighter. When they seized Kabul in September 1996, they banned women from going to work — in a city with at least 50,000 war widows — and a few zealots threatened to hang any female Afghan Red Cross employee who dared appear at the compound. The Taliban also decreed head-to-toe *hijab* and beat violators in the street.

But the apartheid is starkest in Saudi Arabia. Most Saudi homes have one entrance for men, another for women. Women ride in the back of the bus in Riyadh, and enter it through a separate door. Until 1981 a woman couldn't meet her spouse unveiled till after the wedding. Saudi daughters inherit half as much as sons. Amusement parks and ice skating rinks have segregated hours, so families cannot visit together. Saudi banks are so segregated that only female auditors examine women's accounts. Medicine is the sole career where the sexes mix, because, though fundamentalists object to women doctors touching male patients, there aren't enough male physicians to go around.

Yet sexual segregation is not an Islamic invention. It predates the Koran throughout Asia and much of the Mediterranean. Some historians ascribe the origin of the veil to Cyrus of Persia (600?–529 B.C.) and the friezes of Persepolis are notably devoid of women. In Periclean Athens, wives lived in the rear of the house and rarely met visitors. They could not make contracts, inherit husbands' property, or sue in court. Carthaginian and Parthian women wore veils and lived in seclusion. The early Christians felt the allure of women's hair might distract even angels during services and asked women to worship in veils, and St. Paul recommended that women shave their hair rather than expose it in church.

At root, purdah seems designed to keep women chaste before marriage and faithful afterward. It is institutionalized gynophobia. In purest form, it becomes a movable cell for women, and it keeps them in the rear of the home even as they walk the busy sidewalk. Purdah blanks the face and erases not just expression, but beauty.

IV Siren

Siren

Beauty is an ecstasy; it is as simple as hunger.
There is really nothing more to be said
about it.

W. SOMERSET MAUGHAM

Six
6

Constellation of Desire

IN Constantinople around 500 A.D., a professional bear-keeper and his wife had a daughter named Theodora. She grew up near the Hippodrome, site of chariot races and gladiatorial contests, a tough neighborhood filled with sideshow owners, stableboys, and roustabouts. But she was gorgeous and brilliant, and at fifteen or sixteen she gained fame as a witty nude dancer. She took a rich lover and moved with him to Tyre, where he abandoned her. Penniless, she made her way to Alexandria, converted to Christianity, and returned to Constantinople.

There, somehow, she met the future emperor Justinian. He was eighteen years older than she, but he fell wildly in love with her. They married in 523, and four years later Justinian became emperor. Or rather co-emperor, for he relied on Theodora's fiber and political savvy. Partly because of her face, she had vaulted from a circus slum to the summit of the Byzantine empire.

Beauty has changed history. The most obvious and spectacular cases have been rulers' favorites, women like Mme. de Montespan, Mme. de Pompadour, and Eva Perón, and less often

men like Essex, Potemkin, and Godoy. Such individuals reap perhaps the highest rewards for looks, though beauty has been a social
booster-rocket for women from Phryne to Pamela Harriman. But
the boons run much deeper. The attractive do better in romance
overall, and we deem them superior on an array of scores.

Hence, most people pay careful attention to their looks. In
one survey of 30,000 people, only 18 percent of men and 7 percent of women said they were not concerned about appearance
and didn't try to enhance it. Indeed, individuals across the globe
strive to improve their faces with cosmetics, tattoos, nose rings,
beauty marks, skin moisturizers. Sometimes, as with tanning and
cosmetics like kohl, the effort proves ruinous over time.

If Maugham is correct and there is really nothing more to
be said about beauty, someone should tell the research community. In 1974 one classic review listed around 50 published studies on the matter. By 1986, the number had ballooned to over
1,000. And in 1997, a computer search of scholarly articles with
the word "attractiveness" in the title generated over 1,000. In
fact, there is a great deal to be said about beauty. It has always
been one of life's magnificent mysteries, and its secrets may
finally be falling away.

Plato's Wings

When Roberta Flack sings "The First Time Ever I Saw Your
Face," she is recapitulating Plato. Both celebrate love at first
sight, delirious, world-shaking love of a face, whatever the mind
behind it. The love of contour is ancient and of high pedigree.

In Socrates' second speech in the *Phaedrus*, he explains.
When we were on high with the gods, he says, we all saw the
reality of the perfect Forms, such as Beauty. The soul still recol-
lects them. But earthly faces do not remind all people equally

well of these glories. Hence some stunted individuals are simply eager for sex and surrender to its lure.

But one recently returned from heaven has a fresh image of Beauty in his mind, and its simulacra rivet him. When he sees a beautiful face, he shudders and feels a chill of fear as he did above. He gazes rapt at the gorgeous person. Sweating and fever come on. The beauty pours in through his eyes, and warms and waters the place where the soul's wings once grew. Hard scabs had sealed these spots off to keep the wings within, but now the spectacular inrush makes the feather shafts swell and push against the scabs, and it feels like teeth breaking through in a child. The gazer seethes and tingles as the wings try to unfold. Pain vanishes and joy perfuses the soul.

Yet when the beautiful face departs, the openings for the wings seal shut again and lock them in. The feathery stumps throb like an artery and prick at the scab, stinging the entire soul. The pain drives it wild. The lover's soul cannot sleep at night. It pines and dashes about to wherever it may find the beautiful person.

In *The Picture of Dorian Gray* Oscar Wilde released a different aria, one of axiom. Lord Henry Wotton says, "Beauty is a form of Genius — is higher, indeed, than Genius, in that it needs no explanation. It is of the great facts of the world, like sunlight, or spring-time, or the reflection in dark waters of that silver shell we call the moon. It cannot be questioned. It has its divine right of sovereignty. It makes princes of those who have it. You smile? Ah! when you have lost it you won't smile."

Both speeches are absurd and wonderful, and both authors seem to sense their own excess. Scholars unable to credit Socrates' sincerity suggest he is offering an example of rhetoric, of the argument that succeeds by appealing to the listener's bias. And Wilde anticipates derision ("You smile?"). Yet beauty seems made for hyperbole. Freud offered a more precise and objective description—it has "a peculiar, mildly intoxicating quality"—and missed the soul of it.

Beauty is a visual pheromone. No other shape affects us like the face, and in literature it has often sparked a match-quick passion. In Shakespeare and Fletcher's *The Two Noble Kinsmen*, Palamon and Arcite spy a woman from a second-story window, instantly fall in love, and squabble over rights to her, regardless of her character or, indeed, her wishes.

Even more extravagantly, people dote on images of strangers. In the *Thousand and One Nights*, Ibrahim sees a picture of a gorgeous woman in a book, falls in love so deeply that he weeps for her day and night, and eventually sets off to find her. When he does, she turns out to be a "virago among viragoes." Women are equally susceptible. In *Khamseh*, by the twelfth-century Persian poet Nizami, Shirin glimpses a painting of the prince Khosrow hanging from a tree in a meadow. She trembles, hugs the portrait, and becomes lost in love. In Kate Chopin's *The Awakening* (1899), an adolescent Edna keeps a photo of an Edwin Booth–like tragedian on her desk, and kisses it in private. The literature of every continent affords similar examples.

So does real life. The adolescent Marilyn Monroe tacked a photo of Clark Gable to her bedroom wall, and countless girls later did the same with the Beatles and the New Kids on the Block. Alone in Boston in 1904, future film mogul Louis B. Mayer fell in love with a woman in a photo and set out to court her; his marriage to Margaret Shenberg proved long and happy.

Indeed, photos have formed the basis of a marriage industry. Between 1900 and 1924 some 22,000 "picture brides" moved from Japan and Korea to the United States, mainly Hawaii. There, to their shock, they often found their husbands much older than their photos had indicated, and living in squalor. Today, mail-order bride catalogs have become a multimillion-dollar love bazaar. *The Glasnost Girls* shows photos of more than 400 Russian women, and in Hawaii over 1,000 men subscribe to the pioneering *Cherry Blossoms*, which began in 1974. This business is ideal for the Internet, and email-order bride websites like Latin Hearts and Pearls of the Orient are proliferating.

To the extent such catalogs work — and *The Glasnost Girls* claims 10 percent of the women in its files get married — it's through the potency of beauty. Scientists have given men photos and brief descriptions of potential dates, indicating at random that they were calm or anxious, trustworthy or untrustworthy, dependent or independent, modest or boastful. The men overwhelmingly chose the better-looking women as dates. Character was irrelevant.

Looks matter especially at the start of a relationship. Appearance is, of course, the one trait we always see first in a person. And attractive men and women overall are more popular with the opposite sex and date more often.

Researchers wondered if good-looking people seek each other out, and discreetly observed couples at movie theaters, bars, social events. They found that couples generally match each other in attractiveness. Good-looking men and women tend to pair up, and so do homely ones. Moreover, the well-matched relationships seem closer. In one observational study, 60 percent of the matched couples engaged in intimate touching, but only 22 percent of the unmatched ones. Matched couples are also more likely to stay together over a longer span of time. In the United States, Japan, Germany, and Canada, researchers find they are more likely to get married and stay married.

When a pair is unmatched, the less attractive person usually brings a bonus to the relationship. He or she is richer, more loving, more prestigious. There is, it seems, a balance effect, in which one lacking looks can add other weights to the pan.

Hence, good-looking women tend to marry up, like Theodora. One 1973 study found they were more likely to expect longer vacations after they married, and to work fewer hours outside the house. In a 1979 study, the better-looking had more say in the relationship.

Men gain in others' eyes by being seen with beautiful women. Researchers showed subjects photos of a man alone, with a beautiful woman, or with a homely woman. They rated the man with the beauty most highly. In second place was the man alone, and lowest was the man with the homely woman. Observers seem to read a man's success in his woman's looks. Intriguingly, women do not gain by being seen with handsome men.

The tug of beauty is its heart. But like a glowing metropolis, its influence reaches far beyond its perimeter.

The Aura

In *Hippias Major*, Socrates dizzies an ordinary mortal with the challenge of defining *kalos*, a sprawling word that means both "pretty" and "good." The task is impossible, given its fuzzy and multidimensional nature, yet "beautiful" and "good" commonly merge in a word. We see it in German (*schön*), Spanish (*bonita*), Zulu (*-hle*), Hungarian (*szép*), Swahili (*zuri*), Amharic (*mälkam*), and a wide range of other tongues.

The association is not just linguistic. "Beauty and goodness are inseparable," said Thomas Aquinas (1225–1274) and in fact people have trouble keeping them apart. Hence they bestow an array of virtues on the attractive. Good-looking individuals seem more competent, likable, happier, blessed with better lives and personalities. In one experiment, people predicted happier marriages and better jobs for them, and indeed rated them lower on only one count: their caliber as parents. A 1970 study found people consider them more amiable, happy, flexible, pleasure seeking, serious, candid, outspoken, perceptive, confident, assertive, curious, and active. They exert more control over their destiny, subjects felt, while the homely endure the world's caprice.

And though many of these results come from lab studies, they also carry over into real life.

For instance, teachers assume the good-looking are more intelligent. In one study, researchers gave 400 fifth-grade teachers a dossier on an imaginary child. Though each contained the same academic record, the photos differed. Teachers in general felt the cute children were smarter and more popular, had parents more interested in their education, and were more likely to get advanced degrees. The bias flows both ways. Students from second grade through college deem attractive teachers more organized, friendlier, and better overall—and fear ugly ones will assign more homework.

Beauty gilds subjective judgments of all kinds. Men rate awkward essays much higher if they think a good-looking woman has written them, and both men and women judge paintings as better if the artist is a pretty woman. However, one study found women deemed a bad essay worse if the author was a comely female, and did not score the essay higher if the author was a handsome male.

The effect seems to have two strands. Psychologist Diane Berry notes that we distinguish babyish faces like Helena Bonham Carter's from maturely attractive ones like Julia Roberts's. The babyish faces seem more naive, honest, warm, submissive, kind, weak, and sincere. Among males, people view "pretty boys" as more kind and artistic, while ruggedly handsome "Marlboro men" seem strong and masculine. In one case we associate good looks with weakness, in the other with strength. So far, however, most studies have not broken down their results by these types.

Physical appeal can also lead to negative stereotypes. People see the attractive as more prone to have affairs and get divorced. Occasionally the pure fact that they seem better off can work against them. For instance, physicians perceive attractive patients as suffering less pain and needing less care.

How far does the halo extend? In 1991 researchers conducted a meta-analysis of previous studies to find out. They discovered that people rate the attractive as much more socially adept and significantly brighter, more powerful, and better adjusted. But good looks barely nudged up scores on integrity and concern for others. These aspects of morality seem relatively immune to the spell. Even so, people rated the good-looking as worse in only one area: vanity. We read character in the face, and beauty casts a glow over the self.

This phenomenon changes the lives of the beautiful. It begins in the maternity ward. Mothers cuddle and coo more over newborn babies if they are cute, and at home parents punish them less often. Among girls, good looks enhance popularity on the school playground. Social skill is more effective, of course, yet scholars note that beauty brings prestige, so people find value in befriending the attractive.

Evidence suggests teachers give higher grades to the good-looking. Like parents, teachers are more reluctant to chastise them, though some actually single them out for punishment, since they have higher expectations of them.

No occupational group is more aware of the bias toward beauty than psychologists. Even so, psychotherapists give good-looking patients better treatment, studies find, and patients in turn have more confidence in attractive therapists.

A 1987 note in the *Harvard Law Review* urged that judges construe the Rehabilitation Act of 1973 to bar discrimination based on "largely immutable aspects of bodily and facial appearance," which would include beauty. Such an interpretation is unlikely, since it would lead to spectacular wrangles, but the attractive clearly receive preference in personnel decisions.

The good-looking earn more money and hold more prestigious jobs than the average, and the average do better than the unattractive, according to a 1978 study. In fact, one survey of Canadians found the good-looking earned 75 percent more than the ugly. Attractive job applicants for managerial positions win higher rankings than homely ones with identical backgrounds. One study showed that better-looking cadets at West Point tended to graduate with higher military rank.

Some researchers have found no pro-beauty bias in jobs that demand little face-to-face contact with the public. Indeed, employers may actually discriminate against the babyfaced in jobs requiring some shrewdness and leadership, such as loan officer. Looks may ultimately count most in hiring when employers choose from a pool of highly qualified applicants. If all candidates can do the job well, it seems, other criteria come into play.

The good-looking are more successful scofflaws. In one story in the *Thousand and One Nights*, a thief steals a trooper's coin bag. The victim accuses him, but people cry, "No, he's such a handsome youth. He wouldn't steal anything!" Similarly, researchers in one study told subjects a tale of a child's misbehavior, then showed a photo of the little offender. Subjects were more apt to rate the unattractive child as wicked, and the attractive one as good, but out of sorts that day. As adults, the good-looking can shoplift with greater ease, since witnesses are less likely to report them. And when they come before a judge or jury, the benefits continue.

Most studies of courtroom behavior have used mock juries composed of college students. In one, jurors tended to acquit good-looking defendants, but only where evidence was vague. In another, jurors favored the attractive unless they exploited their looks in the crime, as in certain confidence games. In such cases, jurors meted out harsher penalties. Good looks have mattered in simulated rape and sexual harassment cases, where they have helped defendants and hindered victims.

Actual courtroom studies are rarer and more difficult, but they involve trained professionals and real-world impact. In a classic examination of felony cases, good looks did not affect conviction or acquittal, but did lighten sentences. Likewise, a 1991 study of 2,235 defendants in misdemeanor cases found that judges gave the attractive lower fines and bail. The researchers speculated that looks would not have affected the conviction phase here either.

Why not? The answer probably involves the different nature of the two decisions. Judging guilt requires an objective appraisal of facts — did she do it? — while setting the penalty entails a subjective rating of character — how bad was she? Since we tend to view attractive people as "nicer," looks influence the latter more. It resembles the leniency effect with smiles.

How true is the preconception? Are attractive people just the grinning recipients of fantastic largesse or do they inhabit a higher plane? In fact, they do score higher on some scales. Studies show that they are more socially skilled, better adjusted, less shy, and less socially anxious. They are also less lonely and more sexually experienced.

The attractive may be vainer as well, if time logged in at the mirror is an accurate measure. The good-looking stare more at their lineaments, perhaps as smart kids gaze longer at their report cards. The beauty Zuleika Dobson, in Max Beerbohm's extravagant novel of the same name, travels with her great cheval-glass, which always "stood ready to reflect her." The Phantom of the Opera, on the other hand, has no mirror in his living quarters.

The good-looking also seem more aware of their appearance. They are clearly better at conveying positive facial signals, like smiles, though not negative ones. Psychologists speculate that since their faces are the objects of more attention, they try to use them more effectively in conversation. Intriguingly, one study found that good-looking people showed little motivational impairment effect when lying.

However, the stereotype is a mirage on several counts.

For instance, we view the attractive as more intellectually competent than the unattractive, and the effect is especially strong when other information about a person is absent. But a 1995 meta-analysis of 30 studies demonstrated that looks do not correlate with intelligence. We gift the attractive with illusory IQ points.

Do looks affect self-esteem? The surprising answer is: Yes, but not much. One recent meta-analysis found only a slight link between beauty and self-regard. Indeed, another study linked self-esteem to only two variables, a meaningful relationship and fulfilling employment, an echo of Freud's values of *lieben und arbeiten*, love and work.

Subtle pitfalls may make self-esteem more elusive for the good-looking. Psychologists Harold Sigall and John Michaela suggest that attractive people, especially women, can find it harder to accept compliments as genuine. They get so used to flattery that they tend to shrug off authentic praise.

The ultimate question, of course, is whether looks bestow happiness. In 1995 a team of researchers tried to answer the question. They probed the "subjective well-being" of attractive and unattractive college students, defining it as a mix of life satisfaction, positive affect, and lack of negative affect. They found that good looks correlated only slightly with well-being and, once again, only slightly with self-esteem. Future studies, perhaps with different definitions, may yield other results, but as far as we know now, beauty isn't bliss.

In all, the beautiful-is-good finding is extremely robust, replicated endlessly and beyond doubt. Yet most of these studies have an interesting limitation, which we'll return to later.

The opposite of beauty-is-good is the ugliness-is-bad syndrome, and literature teems with it. Of the Phantom of the Opera, the Persian detective says, "His horrible, unparalleled, and repulsive ugliness put him without the pale of humanity." Frankenstein's monster is also hideous, and both turn to evil because of rejection for their looks. As Dorian Gray grows viler, his portrait becomes more repulsive. The Wicked Witch of the West is no beauty contestant.

Even today, most movie heroes and heroines are better-looking than the villains. And studies indicate that people view the ugly as more likely to accept bribes, to have epilepsy, and to be politically radical. It is an unpleasant stereotype, and it can be devastating when it turns in on itself.

Delusions of Ugliness

In Picasso's *Girl before a Mirror* (1932), a blonde maiden stares into a looking glass. There, her face is darker, her forehead is a scarlet stain, and her hair is green. She is gazing at a chimera. Her mind has warped her face.

Studies show that almost everyone dislikes at least some aspect of his or her appearance. For instance, Pearl Buck worried that her nose was hawklike, though few concurred. In one 1972 survey, 23 percent of women and 15 percent of men were dissatisfied with their looks overall. By 1986, fourteen years later, the percentages had jumped to 38 for women and 34 for men. We are growing more critical of our faces.

But some people go much further. Like Picasso's woman at the mirror, they imagine awful blemishes. The clear skin is blotched, the snub nose is swollen, the thick, lustrous hair is sparse, dull, vanishing. They often try to hide this concern, but their actions can baffle friends and family. They may avoid society, shun jobs, even shut themselves indoors. Indeed, their behavior parallels that of the truly disfigured.

Such individuals suffer from an ailment called body dysmorphic disorder, or BDD. Though clinical descriptions of it date back at least a century and it appears in similar form in all cultures, scientists have only recently recognized it as a distinct illness. It may affect as much as 2 percent of the population, or 5 million Americans.

BDD is an obsession. Its victims don't just think they look ugly. The notion preoccupies them. They say things like, "I can't stop thinking about my hair" and, "It's like a second reel always going," and they spend at least an hour a day fretting about their unsightliness.

This problem afflicts the attractive and unattractive almost equally, and in any degree from mild to crippling. Some sufferers realize their worries are unfounded, but say they simply can't help themselves. Others truly believe they have defects, despite all evidence to the contrary. For instance, one woman felt her gorgeous mane was "frizzy and ugly," though friends often asked her stylist's name to try to duplicate it. Another woman deemed herself ugly, yet she'd received requests to work as a model.

Asked to draw self-portraits, these people often include their perceived deformities. One woman depicted her face as a mat of hair and said this picture replicated her image in the mirror. A man obsessed by his nose drew nothing but a nose, indeed, three zucchini-like noses, and placed large circles on them to indicate his pores. In such cases BDD seems to cause ongoing hallucinations.

Sufferers tend to agonize about a single feature, sometimes one beyond the face, such as the legs, hands, breasts, and genitals. They most often fix on the nose, and one study found 45 percent of cases involved it. Many sufferers in Japan worry about their eyelids, a rare concern among non-Asians in the United States. The precise flaw of the afflicted feature also varies. Hair, for instance, can be too flat or too upright, too dark or too light, too thick or too thin. Intriguingly, when asked to rate their looks without the defect, BDD sufferers are rather accurate. One feature hexes the whole.

The mirror fascinates most BDD victims. It holds their fate, their very opinion of themselves, and about 80 percent check mirrors compulsively. They typically want to see if they look as bad as they feel, and the mirror can buoy them up or crush them. Results of a morning mirror-check often set the tenor of the day. When driving, some people gaze into the rearview mirror so fixedly they have accidents. A few, on the other hand, dread mirrors and won't enter places where they abound, like restaurants.

Sufferers attempt many remedies. They may groom themselves excessively, as by brushing their hair for hours. They often wear camouflage: wigs, hats, makeup, skin bronzer, dark glasses, even masks. They may pick endlessly at their skin to rid it of "white spots," "small black dots," "ugly things," and some have poked holes in their cheeks. They may seek surgery, and even when they find a doctor who will perform it, they rarely like the outcome. Occasionally they operate on themselves. One man who thought his fingers too long sliced them off. Another, despairing of his nose, cut it open and tried to replace it with chicken cartilage.

At its worst, BDD can lead to death. One attractive woman commonly ran red lights because she didn't want people gawking at her "ugliness" when she stopped. More often it inspires suicidal thinking, and suicide itself.

What causes this odd ailment? No one knows, and there are many overlapping explanations. Psychological history may play a role. Katharine Phillips of Brown University, the nation's leading authority on BDD, found 60 percent of patients had been teased about their looks when young. The significance of this fact is unclear, since we don't know how many nonsufferers have been teased. But people with body-image problems in general have endured more hazing as children and adolescents. The illness also seems to correlate modestly but positively with trauma, especially sexual trauma.

Personal traits may contribute. A high proportion of BDD sufferers have depression and many report low self-esteem—two qualities of individuals afflicted by other kinds of body-image distortion.* However, it is unclear which causes which, or whether a third factor spawns both. A package of further traits, including social anxiety, shyness, perfectionism, and sensitivity to rejection, also tend to accompany BDD.

*On the other hand, according to psychologists Laura Longe and Richard Ashmore, "people who have high self-regard in general tend to feel physically attractive."

Of course, the brain is the ultimate source of behavior, and so far the most successful treatments for BDD have been biochemical. The compounds known as serotonin-reuptake inhibitors have proved especially effective.

Serotonin is a neurotransmitter, one of the chemicals that conveys a message from one nerve cell to the next by jumping the synapse between them. The arrival of an electric pulse sparks its leap. At the other side it plugs into a receptor like a key in a lock, and triggers reaction in the new neuron. In this way, electrical current can pass selectively through the fabulously interwoven web of possible routes in the brain.

Serotonin affects a bewildering variety of functions, such as sleep, appetite, sex, mood, pain, cognition. It also plays a role in depression and OCD, or obsessive-compulsive disorder, which BDD resembles somewhat. We don't really know how it can be so versatile, though the answer may involve the wide array of serotonin receptors, at least 15 at latest count.

In some patients, medications which reduce serotonin levels lead to delusions and worsen BDD. LSD, in fact, counteracts serotonin and triggers hallucinations, some of which involve distorted body image.

Serotonin-reuptake inhibitors like Prozac increase serotonin in the brain. After taking them, some people report freedom from the goblins of BDD. One woman stated that within a few days of taking Prozac her facial hair had gone. She was emphatic. She no longer saw it, though she couldn't explain why it might have vanished. Later, when she reduced her dosage of Prozac, the hair returned. Another man said the holes in his teeth shrank and vanished when he took clomipramine (Anafril), but opened up again when he lowered the dosage.

The chemical dopamine, which plays a role in delusional thinking, may also be involved in BDD, and may act in concert with serotonin. Both, for instance, are involved in OCD. Medications that counteract dopamine help alleviate the symptoms of OCD.

BDD is so debilitating largely because it affects our whole sense of our looks. Everyone is vulnerable here, so peoples across the globe have always enhanced their faces.

Makeup Babylon

Marilyn Monroe was obsessed by her face. As an adolescent she washed it up to fifteen times a day, and she later became notorious for spending eons before the mirror, redoing the makeup around her eyes or mouth. Some thought her bewitched by her own looks, and a studio executive once snapped, "Hasn't that bitch left her dressing room yet?!" But in fact she was creating her beauty.

Makeup can have profound effects, as her career illustrates. Some anthropologists suggest it originated with bug repellent, the lotions peoples like the Northwest Indians smeared on their skin to ward off gnats and blackflies. It may also have begun as a wind or sun guard.

Even these lowly purposes can yield an esthetic. The Tiv of Africa coat their bodies with anything that will yield a sheen: vaseline, castor oil, palm oil, groundnut oil, a paste made from camwood. Such lubing makes the skin glow, and glow is a principle of beauty among the Tiv.

The step from patina to pattern was short. When Western explorers sailed out to reconnoiter the world, they found it scattered with primitives who decorated their faces and in Europe the painted face came to epitomize savagery. But the scorn flowed both ways, as it usually does. In Brazil, Eyiguayegui Indians told missionaries they were stupid for not wearing paint. It made one a man; without it, one was no better than a beast.

Most tribal peoples adorned their faces with abstract art. In *Tristes Tropiques*, Claude Lévi-Strauss describes the Caduveo women of Brazil, who spun out complex arabesques on their faces. This tracery reminded him of the wrought-iron work and stuccos of Spanish baroque, and it enchanted him.

Caduveo maidens wore makeup for basic human dignity, as well as for status and erotic ends. But Lévi-Strauss believed that, ultimately, their designs expressed an unusual "collective dream." The facial patterns were "hieroglyphics describing an inaccessible golden age, which they extol in their ornamentation, since they have no code in which to express it." Lévi-Strauss was the leading exponent of structuralism — the idea that psychological universals emerge in cultural structures — and this remarkable passage suggests that hidden sociopolitical yearnings surfaced in their makeup.

In the West, makeup has been legerdemain. Like fashion photography, it breeds beauty by shifting patterns of light on the face. Egyptians first donned cosmetics to block the hot Sahara sun and quickly saw their value in allure. Ladies favored green malachite for eye shadow, as well as black antimony and kohl. Men wore paint; Tutankhamen's tomb contained a wealth of makeup vials. Priests took cosmetics beyond facial illusion to all-purpose magic. The proper body marks invoked the gods, and to please a deity, the supplicant applied a precise color and shape, often while performing a sacred ritual.

When Cleopatra sailed her golden, perfumed barge up the Cydnus to meet Antony, she was almost certainly wearing exotic makeup. She typically daubed blue on her upper eyelid and malachite green on her lower, a peacock combination. She lined her eyelids and darkened her eyebrows with black kohl, and smeared white ceruse over her face, neck, and breasts. She colored her cheeks with yellow ochre and her lips with a carmine dye. Tradition credits her with a treatise on cosmetics.

Roman women learned makeup from the depraved Egyptians, and some Roman men favored it as well. Nero, for instance, applied ceruse and chalk to whiten his face, kohl to darken his eyes and lashes, and red fucus to redden his cheeks and lips. In *The Art of Beauty*, Ovid offers recipes for making cosmetics, one of which includes white lead, and in Augustan England people tried to follow them.

In the Middle Ages, priests frowned on makeup and most faces were bare. But the Crusaders brought back perfumed oils, face whiteners, and hair and lip dyes from the Middle East, and they revived cosmetics. Harlots seized on them first, but soon great ladies became adepts of the face. The glass mirror enabled precision cosmetics, and the rise of portraiture in the Renaissance popularized it further.

By the time of Elizabeth I, white ceruse had become the most popular cheek-blancher. One mixed white lead on a palette, often with vinegar and egg whites, then applied this paste to the skin with a wet cloth. It created a spectral visage.

As late as the 1950s, some dermatologists believed the skin an impregnable shield. In fact, it is a three-dimensional landscape, webbed with tiny rifts and stippled with pores, and some chemicals slip right through it. That's why we can die from touching certain tropical tree frogs, and why sarin, the nerve gas unleashed on a Tokyo subway line, can kill by contact. Other substances harm more slowly. The pesticide DBCP percolates through the skin and causes sterility in men, and benzene induces leukemia.

Many ancient cosmetics were also slow killers. White ceruse was poisonous. It contained lead oxide, lead carbonate, and lead hydroxide, all of which seeped into the skin and accumulated. By the end of Elizabeth's life it had ravaged her face. She banned mirrors from the palace and thus, according to Ben Jonson, mischievous court ladies sometimes discreetly daubed rouge on the tip of her nose. In the early 1600s, wreckage to ladies' faces helped spur a vogue in masks. Ceruse apparently killed society belle Maria Gunning in 1760 and actress–courtesan Kitty Fisher in 1767.

Kohl, Cleopatra's famous eye shadow, is mostly lead and antimony sulfides, another harsh brew. Fashionable women dyed their lips with red fucus, made from mercury sulfide, and it sent many of them to the grave. They dyed their hair golden with arsenic and ran lead combs through it to slowly darken it; both treatments were toxic. To extirpate freckles, they commonly used Soliman's Water. Its main ingredient, sublimate of mercury, not only dispatched the freckles, but peeled away the outer layer of skin, etched into the flesh, and ultimately poisoned its victims.

Other cosmetics also had sinister consequences. In the seventeenth century women put belladonna in their eyes to dilate the pupil and achieve a fetching black gleam. This treatment made the eye unable to resist glare and led to glaucoma. To beautify their hair, some women rinsed it with sulfuric acid, alum water, and oxalic acid, substances which made them go bald. They often brushed their teeth with pumice powder, which stripped away the enamel. The quest for beauty could corrode a face by thirty.

Some women sensed the danger. In Firenzuola's *On the Beauty of Women* (1548), Mona Lampiada says cosmetics ravage the skin, ruin teeth, and make women resemble "carnival clowns all year round." Think of Mona Betola Gagliana, she says, who "looks like a gold ducat that has been dropped in acid." By the late eighteenth century, deadly cosmetics had turned British women away from makeup, and by the accession of Victoria in 1837 it had almost vanished.

The sentiment against cosmetics was so strong that, to spur sales near the end of Victoria's reign, entrepreneurs like Helena Rubinstein (1870–1965) had to market them as health aids. Rubinstein, who immigrated from Poland in the 1890s with twelve pots of her mother's beauty lotion, described her first "Valaze" cream as a skin cleanser. Elizabeth Arden (Florence Nightingale Graham, 1878–1966), a Canadian truck farmer's daughter who renamed herself after "Enoch Arden," opened her first salon in 1909 and offered products like Venetian Cleansing Cream and Venetian Pore Cream for blackheads. This approach bred empires.

Cosmetics do have at least one health virtue: They block sunlight and thus reduce carcinomas. Lipstick protects against lip cancer, which men are seven times more likely to get than women. Indeed, some dermatologists have recommended that men regularly wear sunblock on their lips.

Not only could the yen for beauty destroy beauty, but in a double irony, it often offended men, who did not relish their role as targets of deceit. The complaint dates back to the Roman Empire, and doubtless further. Propertius tells his Cynthia, "No beauty parlor for you, believe me:/ Naked Love loves no artificial beauty." Hamlet says, "God hath given you one face, and you make yourselves another." The critics are legion and include Lucretius, Martial, Juvenal, Robert Burton, the first Samuel Butler, John Evelyn, Jonathan Swift, and ranks and ranks out to the horizon. In 1770 the English Parliament passed an act voiding marriages into which women had lured men by cosmetics. This comic legislation went unenforced.

But male advisers to women have typically urged makeup. The *Kama Sutra* of Vatsyayana (c. 2nd cen. B.C.) told women to tantalize their men with cosmetics, and also recommended tattooing and staining the teeth and hair. "There's nothing amiss in darkening eyes with mascara," says Ovid. "With sufficient neglect, Venus would look like a hag." But he also warns against excess: "Who wants to look at a face so smeared with paint that it's dripping,/ Oozing sluggishly down to the neck of the gown?" Of course, men rarely object to what they don't see. As John Donne wrote, "What thou lovest in her face is colour, and painting gives that, but thou hatest it, not because it is, but because thou knowest it."

The literati have checked in. But do men overall dislike makeup? Psychologist Don Osborn of Bellarmine College tried to find out. He showed 50 male and 50 female raters photos of women with and without makeup. No matter how good-looking the female subject originally, makeup boosted her attractiveness ratings by more than one standard deviation. This huge jump spurred Osborn to urge that all women consult a professional makeup artist for advice on the best use of cosmetics. Moreover, makeup hiked scores even when raters were aware of it. Osborn suggested that cosmetics also act as a signal, a pleasing cue that the woman is receptive. Even so, about 20 percent of raters preferred the unmade-up models.

Of the Caduveo, Claude Lévi-Strauss said, "The erotic effect of facial makeup has never been more systematically and

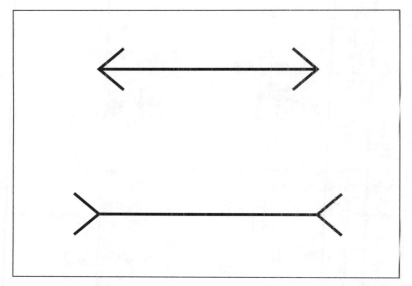

The Müller-Lyer Illusion.

consciously exploited." The anthropologist had never heard of Hollywood. Americans pay some $30 billion a year for cosmetics, and the summit of the art lies in film studios. Not only do they hire top professionals, but makeup can be bolder onstage than in the street.

In the well-known Müller-Lyer illusion, a straight line with flanges pointed outward seems longer than one with them turned inward. The one tugs the attention beyond the line and seems to lengthen it; the other pulls it centerward and shortens it. Makeup artists exploit this effect. For instance, by daubing shadow above the outer ends of the eyes, they draw them farther apart. By applying it above their inner corners, they narrow them. Eyebrows too can stretch or shrink the zone between the eyes. When they rise high and away like birds taking flight, the orbs seem more widely spaced. When they slope close to each other like a funnel, the eyes draw together.

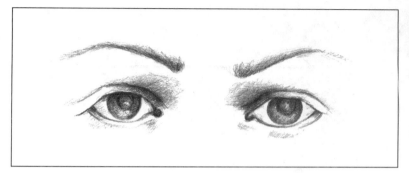

Narrowing with Eye Shadow.

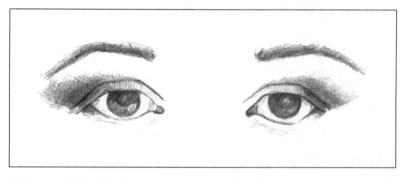

Widening with Eye Shadow.

Shadow subtracts. It suggests hollow, absence where there is flesh. It can thus seem to carve away parts of the face. For instance, one can narrow the nose by applying shadow to its sides, and shorten it by putting a dark color below and blending it up over the tip. Makeup artists minimize a long chin by placing shadow on it, and treat a double chin the same way.

Highlight was once white or yellow greasepaint, and now is any color opposite that of shadow. It emphasizes, makes a feature stand out. Apply it over the tip of the nose and the nose will seem longer. Daub it on a short chin and it too grows. And a band of highlight run down the center of the nose will widen it.

The eyes are a vital hub for makeup, as most women realize. Eye shadow helps enlarge the orbs, and goes on the upper lid only, since if placed beneath, it suggests bags. Some makeup artists add a bit of blush below the outer end of the eyebrow to give the eye sparkle.

Eye lining, applied with a brush or sharp eyebrow pencil, is a slender blur along the lids next to the lashes. On the upper lid it usually begins a third of the way out from the nose and extends to a quarter of an inch beyond the eye, where it ends in a curve. A similar line on the lower lid begins two-thirds of the way out. Makeup artists stress that the blurs should fade away rather than end abruptly, and that they should not meet.

Rouge or blush stands out, and most women use it on the cheekbones, to enlarge them. Hollywood makeup artist Richard Corson says women should never put blush closer to the nose than an imaginary line dropping from the center of the eye. Indeed, a woman with a narrow face should place it further away, to take advantage of the Müller-Lyer illusion. A broad-faced woman should avoid getting it too close to the ears, and indeed should apply it as a vertical smudge.

Paint can redefine the apparent shape of lips, since it is more conspicuous. Thus lip color reaching above and below the lips can compensate for narrowness. When the lips are too thick, the task is harder, though makeup artists will often color just the inner part of the lips. Lip color can similarly shorten a wide mouth and extend a short one.

Cosmetics are transient, almost by definition. They can suit an occasion. But other face alterations must suit a life.

The Frescoed Face

The face seems a sanctum. We don't trail indelible decorations over it. Even the Hell's Angels and *yakuza*, men who tattoo bleeding corpses on their biceps, have honored the face. It is tempting to think the face is so basic we shirk from toying with it. It teems with meaning, and we'd no more drill ink into it than paint a TV screen.

It isn't true. Face tattoos are ancient and widespread. They go back at least to the late Neolithic, when farmers adorned their faces with blue tridents. The ancient Thracians, Assyrians, and Britons all tattooed their faces.

But Leviticus 19:28 forbade scarification and tattoos: "Ye shall not make any cuttings in your flesh for the dead, nor print any marks upon you." The Emperor Constantine commanded that no one tattoo the head, because God made us all in his image, and in 787 the Second Council of Nicaea banned tattoos. Such strictures made them almost vanish in Europe, except on wild fringes like England. After the Battle of Hastings, legend says, Harold's queen wandered the corpse-strewn field until she found his mutilated body, which she recognized by its tattoos, including her own name over his heart.

Face tattoos have abounded elsewhere. They appear in northern Africa, the Middle East, and India, and among the native cultures of North and South America. For instance, the Jamestown settlers of 1606 found Indians with animals and snakes tattooed on their countenances.

Tattoos are a signal system and convey a variety of messages. Among men, they classically flash a warning: Don't trifle with me. Women have generally worn them for beauty, though men have also exploited their esthetic appeal. Both sexes have used them to convey ancestry, group membership, and prestige.

A face tattoo could be both stigma and trophy. In ancient times it often stamped the outcast. The Romans tattooed slaves between the eyes, Burmese rulers drilled a circle on the cheeks of habitual criminals, and Hawaiians branded pariahs, or *kaauwa*, with facial tattoos. In early Japan, face tattoos were a criminal punishment. The *Nihon Shoki* (c. 720 A.D.), the semi-mythic saga of ancient Japan, tells of a rebel whose life the emperor spared, tattooing him near the eye corner instead.

But male face tattoos have also recorded triumphs, like notches on a gun. Among Inuit of Victoria Island, a man who slew a whale received a line from his mouth to his earlobes, and one who killed a warrior earned two lines from nose to ears. Inuit debated which deed was greater. Some peoples tattooed a mark on the face for every human head they took. A voyager to Indonesia saw a man with twenty-nine, and one old Melanesian chief sported ninety-five. Such men wore their homicides on their faces.

Females have commonly used face tattoos as decoration, especially on the chin. In Dangs, one of the poorest areas of India, Kumbi and Warli women tattoo dots and lotus symbols on their faces, both for beauty and as skeleton keys to heaven. Ainu women tattooed the skin around the lips, often making them seem huge and swollen.

Among Alaskan natives, women received chin tattoos at puberty, usually one to five vertical lines. They flagged nubility. When a young girl begged for a tattoo, elders sometimes warned her that, once she got it, men might rape her. On Kodiak Island, women tattooed their chins, and *schopans*, boys raised to serve as "women" and catamites, received chin tattoos as well. According-ing to one missionary, Inuit tattooed girls to prevent the evil spirits of the afterlife from turning them into drip-containers for seal-oil lamps.

★   ★   ★

"Tattoo" is a Polynesian word, deriving from *tatau* ("to strike") in Samoan and Tahitian, and face tattoos reached apex in the South Pacific. They were most common among the two fiercest peoples: the Maori in the southwest of Polynesia and the Marquesans in the northeast.

When Captain Cook reached New Zealand on his first voyage in 1769, he found Maori with the colorful face tattoos called *moko*. Moko were curvilinear, often spirals and eddies in a dizzying swirl. Males commonly tattooed the entire face, from hairline to throat. They left a strip of untattooed skin 1/8 inch wide called the *waiora* down the center of the face, like a fold.

Moko made the face look fearsome. Tattoos near the brows and the mouth exaggerated Maori battle grimaces and presumably unnerved the other side. They may also have disrupted the face-recognition system, so the warriors seemed less human. Indeed, the whole-face tattoo is chromatically feral. It changes the face like no other adornment on earth.

Maori women usually inked just the chin and upper lip, leaving the rest of the face clear. They incised curving vertical lines and semicircles on the chin, and black horizontal stripes on the upper lip, apparently because the Maori found red lips offensive. Now and then women etched designs into the nose and the forehead just above it. These inkstains were central to the lives of women too. When the firstborn daughter of an important chief received her tattoos, the Maori often celebrated with a human sacrifice.

Both Maori men and women viewed tattoos as glamorous and erotic. The female chin tattoo was called *whakatehe*, from *tehe*, or "erection," and the scholar Alfred Gell calls it "a representation of the phallus probing towards its predestined (open) orifice." Hence Maori tattooers refused to touch *puhi*, the high-ranking vestal virgins.

Each moko was unique. On the brig *Active* in 1815, the British presented the chief Themoranga with his first pen, and he at once drew his moko from memory. Maori signed land contracts with Europeans by drawing their moko, and face tattoos served as signatures for chiefs on the 1840 Treaty of Waitangi, which brought New Zealand into the British Commonwealth.

Like most primitive tattooing, the Maori operation was painful. The tattoo artist dipped a comblike bone instrument in dye and hammered it into the subject's flesh, creating a line of stained punctures. He then shifted it and drilled again. Because of the pain, large designs required many sessions. Moreover, some people tattooed highly sensitive areas like the eyelid, tongue, penis, and labia. The agony, naturally, enhanced the status of the men and women who endured it.

Powerful taboos surrounded this procedure, partly because of the presence of blood, and afterward Maori destroyed the hut where it had taken place. Inflammation throbbed around the tattoo for several weeks after. Pus often oozed out, as infection was a constant threat and sometimes killed the tattooee.

Moko survived death. Maori kept the tattooed heads of male kin and foes, called *moko makai*, and smoked them to preserve them. They pried open the lips of enemies, sewed them into a gape, and mocked them. They also traded such heads for the *moko makai* of their own blood, whose lips they promptly sewed shut. The latter were intensely taboo, and Maori guarded them carefully.

Western wealth vaporized this custom. On Captain Cook's first visit, his naturalist Joseph Banks bought a head for a European museum, and soon a lively trade sprang up. Crewmen of whalers purchased heads, often with rifles, and the customs house in Sydney listed "Baked Heads" as a regular entry. The commerce continued until 1831, when the Australian government banned it. Near the end, when the *moko makai* were running out and the Maori were desperate for rifles, they often hastily tattooed the face of a slave and beheaded him. They also tattooed the faces of the dead, a practice which degraded product quality since it left no ink in the lower skin layers. Europeans didn't seem to notice. Today, many Maori are pressing museums for the return of heads which, because of their moko, they can identify as specific ancestors killed over 200 years ago.*

In New Zealand, moko declined drastically, through fashion and missionary zeal, and had almost vanished by the wars of the 1860s. In 1921 one observer mourned that only seven Maori males still had them, though hundreds of women did. Today Maori urban gangs are reviving the practice. In Lee Tamahori's film *Once Were Warriors* (1994), the Hekes's angry son Nig joins a gang and gets a flamelike blazon on half his face. Much gang tattooing takes place in jail at night, out of sight of guards. Gang members use smuggled tools and shared needles, risking hepatitis, tuberculosis, blood poisoning, and HIV.

---

*The Jivaro Indians of Ecuador long specialized in head-shrinking, and their tiny trophies—or *tsantsas*—also became hot articles of commerce. To make them, Jivaro headhunters removed the skull and flung it into the river as a gift to the anaconda, then boiled the head. This process alone reduced it by half. Over the next six days they repeatedly inserted hot pebbles into the empty cavity and left them there to cool. Occasionally they massaged the head to help it dry and keep its shape. When it had shrunk to the size of a fist, they sewed up the mouth to prevent escape of the spirit. A five-day saturnalia then ensued, fueled by manioc beer and spiced by assignations in the nearby forest. One informant told anthropologist Michael Harner, "The desire of the Jivaro for heads is like the desire of the whites for gold." Tourists still avidly seek these shrunken faces, though today most are fakes, made from monkeys.

Queequeg is almost certainly a Marquesan. It's clear from his geometrical face tattoos, which startle Ishmael at the Spouter Inn. "Good heavens! what a sight! Such a face!" Ishmael shudders as he gawks at the large, dark squares. While Maori males wore tattoos mainly on the face, thighs, and rear. Marquesans like Queequeg might sport all-body tattoos, which they called *pahu tiki*, "wrapping in images."

For Marquesan males, tattoos were armor, protective skins. The most common face pattern was three horizontal bars of ink. They formed a dark backdrop for eyes and mouth and heightened a warrior's expression of fury. Spiral tattoos adorned some Marquesan eyes, a further attempt to distract and rattle the foe.

Westerners tend to view tattoos as self-defilement. Is it just cultural bias? The answer is unclear, but in Polynesia tattoos also bore a taint. In much of the region, newborns arrived from *po*, the lands of gods and sanctity. They emitted powerful *tapu*, a menace to others' bodies, and society had to desanctify them. Tattooing sufficed. Thus certain high chiefs never wore tattoos, to retain their sanctity on earth.

The Marquesans believed in an elysian afterlife ruled by the goddess Oupu, whom tattoos disgusted. She placed fierce gods at the portals who ripped apart any new arrival with the hint of a tattoo. Hence the wife or relatives of a dead man scrubbed his corpse for months to extirpate the ink. During this time, the body lay in sacred ground, and afterward it left by canoe for paradise.

In addition to tattooing the face, Polynesians often tattooed faces on the rest of the body. Faces have eyes, and eyes protected one from the *tapu* presence of other bodies. In Tahiti, a common form of assassination was the club blow to the back of the head, which an extra pair of eyes might deter.

In the nineteenth century, priests drove the tattoo out of Polynesia, decrying it as barbarous. Polynesians first omitted tattoos from the face and hands, the most visible body parts. Soon many island cultures outlawed the art, as Hawaii did around 1830. Occasional rebellions revived it, and during the Maori Wars of the 1860s, for instance, moko bounced back as a statement of cultural identity. But by 1890 tattoos had vanished from most of Polynesia except Samoa.

At the same time they were spreading across the underbelly of the West. British sailors discovered tattoos in Polynesia, where they quickly noticed that, especially in Samoa, Tonga, and the Marquesas, women ridiculed clear-skinned men and refused to sleep with them. Most *Bounty* mutineers got tattoos. Sailors brought these exotic markings back from half a world away and soon tattoo parlors sprang up along the waterfront and skid row. Sporadically, more prominent people got them, like King Edward VII, Czar Nicholas II, Lady Randolph Churchill, and eventually Roseanne and Johnny Depp. Lenin reportedly had a skull tattooed on his chest, and Stalin a red star. Churchill wore his coat of arms. In the 1980s punks tattooed their heads, arms, and legs with skulls and swastikas.

Early Hawaiians often tattooed their tongues in remembrance of the dead. The operation was especially bloody and painful, but they endured it because, they said, the gesture was permanent. Today Westerners flock to tattoo parlors for exactly the opposite reason. Laser-wielding surgeons can now extirpate the stale picture, the outdated symbol, the fiery profession to a departed love. So tattooing has become a multimillion-dollar industry, one that includes artists, equipment makers, doctors, and journalists for magazines like *Tattoo*. Customers must still take care, but the timing is good. The tattoo signaled recklessness in the past, and the aura lingers.

In the West, tattoos have occasionally been makeup. In the early 1900s, famed tattooist George Burchett specialized in tinting ladies' cheeks, reddening lips, and darkening eyebrows. After World War I, he tattooed normal skin tones onto the wounds of veterans. But few tattooists had Burchett's skill, and botched jobs proved a nightmare.

The laser is also spurring a mini-boom in cosmetic tattoos. They don't run with perspiration under hot lights, so rock stars have gotten them, and soap-opera actresses have tattooed their lips so they needn't apply lipstick after each take of a kissing scene. And they're reaching the broader public, since they eliminate the rote of applying and removing eyeliner, eyebrow color, and lipstick.

But the problem of art remains. Cosmetic tattoos must be precise and subtle or the backfire is loud, and clumsy work has swelled the appointment books at the removal offices.

And tattoo removal, though surgically simple, is far from perfect. It takes time, usually from four to twelve treatments, with six weeks between each. Moreover, ink erasure can expose scars from tattooing, and lasers themselves can scar, though the risk is slight today. In a few cases the skin darkens or lightens at the site, and some designs, especially multicolored ones, linger on in ghostly form. Tattoos are not yet refrigerator decals.

The Disk and the Ring

"I could hardly have believed it had I not seen it," wrote explorer Jean-François de la Pérouse after landing in British Columbia in 1786. He had discovered Tlingit women wearing one of the world's most remarkable ornaments: the lip disk, or labret.

These women slit their lower lips and inserted bowl-like plugs of wood, stone, bone, or ivory, which pushed the lip out three or four inches. It could extend further. One later trader found an old woman with a labret so large she could flip it up and almost hide her face. The custom was, La Pérouse thought, "the most disgusting perhaps that exists on the face of the earth."

He urged them to shed the adornment. "They agreed to it reluctantly," he wrote. "They made the same gestures and showed the same embarrassment which a European woman would show at baring her breast." Once emptied, their lower lips dropped to the chin and, La Pérouse added, "this second picture was little better than the first."

The custom was noisome in other ways. Lip-plugged women could not close their mouths, and brown tobacco juice trickled constantly from their lips. The plug sometimes fell out when they ate or drank. When women stuffed their mouths too full, they often removed the labret and placed a wad of half-chewed food on it, a practice that offended even tough sea captains.

But the Tlingit considered the labret a mark of special dignity. Only free Tlingit could wear it; slaves could not. And women of highest status wore the largest plugs, and might steer canoes.

The labret was a lifetime's creation. The Tlingit slit the lip of girls as young as three months, added a copper plug to keep it open, and widened it over time. The ritual of inserting the disk occurred at menarche, amid feasting and merriment, since lip plugs indicated a girl's nubility. Throughout adulthood, women further dilated the lower lip by inserting bigger and bigger labrets.

The advent of Europeans caused the decline of labrets, and when German geologist Aurel Krause reached Tlingit lands in 1881, he found only one old woman wearing this ornament, which he said fully warranted earlier assessments of it. But he also noticed that many women had modified the device. Instead of lip plugs, they inserted silver pins or pegs of bone or ivory into their lower lips. As late as 1889 among the Aleut and Chugach, one observer saw lip plugs made of coal and glass bottle stoppers.

In many places, only men wore labrets, as among Inuit north of the Yukon River. The heavy plugs exposed teeth and gums, and Inuit had to remove them in subzero weather to keep their lips from freezing. Aztec men had gold labrets, and the priest who cut out beating human hearts for the gods wore an azure lip disk. A seventeenth-century missionary in Mexico said the "Acaxee" Indians slit the lower lip of a murderer and inserted a bone from his victim as a labret.

Botokudo men of Brazil wore labrets so big they reached their breasts. A man who removed the lip disk seemed to have two mouths, one atop the other, a sight that startled Europeans. (The Botokudo received their European name from the Portuguese term *botoque*, a plug or stopper.) The Suya, another Brazilian tribe, believed their red lip disks marked them off as different from any other people on earth, and that therefore no other group was fully human.

Aside from the New World, lip plugs mainly flourished in Central Africa. Among the Fali, an African society where women wear labrets, mothers hand down knowledge to their daughters that ultimately stems from the precepts of an ancient frog. Labrets give force to the teaching by making their lips look like frogs'.

Labrets show the difficulty of separating beauty from prestige. Both enhance allure. Did Tlingit maidens think lip disks made them more attractive physically or socially? The fast decline of labrets after contact with Westerners suggests the latter.

Hang a jewel from the earlobe, Agnolo Firenzuola says, and not only does the ear not lose, "but on the contrary gains from it and the jewel suffers the loss." Earrings are ancient and rarely controversial. In the Mesopotamian myth *The Descent of Ishtar*, the goddess of love wears earrings as she reaches the underworld, where the gatekeeper tells her to strip them off as a matter of protocol. Hell is too dainty for earrings.

Earrings were popular in antiquity, partly because women often kept their hair pulled away from the ears, so their ornaments were visible. Etruscan women wore a tube-like hoop with the head of a lady, river god, lion, or ram, and Greek women dangled disks and tiny statues of Eros from their lobes. In western Europe between the eleventh and the late sixteenth centuries, elaborate hairstyles hid the ear and earrings almost vanished. They returned in spate in the 1600s, and the best-known form debuted around 1660: the girandole. It is a surmount, often a stylized bow, from which three pear-like drops hang. Anne of Austria owned many girandoles, some with enormous diamonds, and the form remains a standard.

Throughout the centuries people have pierced the lobe for earrings. But the screw fitting, developed at the end of the nineteenth century, become widely popular in the 1920s, as people came to view piercing as savagery. The screw fitting could not hold heavy earrings, but the earclip, introduced in the 1930s, could, and it dominated fashion for the next forty years. Piercing returned in the late 1970s, its barbarism now an exotic tang, since pierced ears are safer and more comfortable with weighty ornaments.

Hence the seventies saw a vogue for great hoop earrings. How big should they be? Ovid warns women against heavy adornments: "Do not burden your ears with precious stones from the Indies,/ Lifted by dusky men out of the watery green." But most women then and now have sighed for such burdens. And Daggoo, the six-foot-five African on the *Pequod*, wore in his ears "two golden hoops, so large that the sailors called them ring-bolts, and would talk of securing the top-sail halyards to them."

In the 1860s and 1870s, women favored novelty earrings which showed hammers, hummingbirds, Brazilian beetles, shovels, buckets, and other amusing items. Lobe ornaments have been organic. Florida Indians pierced their ears and inserted small fish bladders, dyed ruby red, which shone when inflated. The Powhatan of Virginia bored two or three wide holes in their ears and wore turkey legs and squirrel claws. Some hung dead rats by the tail. Others fastened live snakes in their ear holes. The green-and-yellow ophidians writhed and occasionally kissed their captors on the lips. Americans who think Dennis Rodman extreme do not understand the possibilities.

But earrings typically feature metal or gems, objects that glitter. Georg Simmel held that adornment acts as a kind of radiance, expanding one's sphere of significance. Hence the yen for precious metals and jewels. The sparkling gem both catches the eye and, he says, resembles a pupil's gleam. In a sense, it is an extra eye.

Simmel also believed that true elegance lay in a stylized, near-abstract realm, which metal and gems epitomized. Jewelry adds a touch of the "super-individual," and metal jewelry "is always new; in untouchable coolness, it stands above the singularity and destiny of the wearer." It is everlasting, a genie serving a mortal, and we borrow it for a flicker of time.

Earlobe extension is a much rarer custom. The Egyptians of the New Kingdom (1559–1085 B.C.) practiced it. So did the Easter Islanders, as the *moai* show, as well as the Creeks, Cherokee, and Chickasaw of the southeastern United States. Melanesians, especially around New Britain, extended the ears of women only. The basic method was to slit the earlobe, then insert larger and larger plugs till it drooped dramatically. Among the Cherokee, the operation caused terrible pain, rendering the patient unable to lie on one side for forty days, so the Indians worked on one ear at a time. At Chickasaw drinking matches, a man sometimes grabbed the brittle loop of another man's ear and yanked it off. Easter Islanders fastened their slit lobes to the top of their ears before battle, like buttoning, to avoid a similar fate.

The lip disk is a curiosity and the earring can be a joy. But what about other body piercings, like nose rings? Do they stand beyond the individual in Simmel's "untouchable coolness"?

They are certainly ancient. An Olmec cave painting at Oxtotitlan, probably dating from before 1500 B.C., shows a lord with a jade nose ring. Seminole Indian women perforated their ears many times along the upper edge. The Creeks and Cherokee pierced the septum and dangled silver rings from it, and the Tlingit hung rings from the septum which hindered cleaning of the nose and sometimes even covered the mouth.

Like tattoos, piercings could record noble deeds. According to Peter Carder, a sailor with Sir Francis Drake whom Indians captured on the north shore of the Rio de la Plata, every time an Indian killed an enemy, he pierced a hole in his face, starting with his lower lip and working up to his cheek, eyebrow, and ears. Tlingit likewise perforated the rim of the ear after a triumph, such as staging a splendid feast in memory of a hero.

In "Achates McNeil" by T. Coraghessan Boyle, the college student Victoria wears a nose ring. "It's a real pain in the ass," she admits. "I suffer it all for beauty." And the irony in her remark suggests the other forces at work.

The nineties have seen a spectacular rise in body piercing, especially among the young. Beringed celebrities are common. Actress Lisa Bonet and Guns N' Roses guitarist Slash wear nose rings. Triathlete Paula Newby-Frazier sports several rings on the rim of each ear, and Olympic sprinter Dennis Mitchell wears a ring through his eyebrow. Drummer Tommy Lee of Motley Crue seems to have latched metal onto much of his body. Young adults and even high schoolers pierce their noses, lips, tongue, navel, nipples, and genitals.

These acts have social meaning. They show an ability to endure pain. They warn. They link one with others. Piercees justify the act on numerous grounds, such as decoration, curiosity, rite of passage, and, in some cases, submission to the will of a sex partner.

Like tattooing or any bodily invasion, piercing brings health risks. Most commonly, inept piercing causes keloids — thick scars that grow — and infection. The outer rim of the ear has cartilage, as does the front of the nose. Piercers aim for soft tissue, but an amateur may cut this cartilage. If infected, scars may replace it, giving rise to anti-ornaments like "cauliflower ear."

The simple presence of metal in flesh can also cause problems. Sometimes a stud becomes impacted and needs surgical removal. Teeth can break if they bite down on lip or tongue rings. In some cases, mouth jewelry has caused drooling, speech impediments, blood clots, numbness, and eating disorders. The field remains largely unregulated, to the dismay of professional piercers, who fear a backlash caused by bunglers.

An ornament can be exquisite or excruciating, but the greatest jewel of all is both silken and natural: the skin.

The Complexion Quest

We are born with brains, muscles, and skull incompletely formed, but our skin begins at acme. Babies churn out a new skin every seven days, so it looks preternaturally fresh, but a seventy-year-old needs four to six weeks. Almost no one likes this slowdown, and many try to resist it. The long journey of the complexion has many milestones, and people commonly fight them as well.

Moles appear very early. They are usually small, dark circles, flat or raised like a tell. Clinically, they are simply clumps of similar cells on the skin, so we can have moles of fat, blood vessels, even hair. For instance, blood moles, called hemangiomas, can appear on the chest as we age. They look like cherries, and some physicians find them beautiful.

But "mole" usually means a pigment mole. We call them birthmarks when we come into the world with them, though true birthmarks are rare. Most moles arise later, probably under a genetic prod, since sometimes everyone in a family has the same mole. By age twenty most of our moles have arrived, and by forty all usually have. The average person has forty moles, most on the central body and upper arms and legs. A mole survives about half a century, so people commonly outlive their birthmarks, and nonagenarians have few moles at all.

A mole lures attention, so its location matters. At the tip of the nose it is worrisome, but on the cheek it can be fetching. In the Roman Empire, and again in Europe starting in the 1600s, women painted beauty marks on their faces. These little mole-mimics highlighted pale skin — Samuel Pepys (1633–1703) found his wife entrancing with her black patches — and also hid small-pox scars and other blemishes. They soon proliferated and jumped the gender line. Fops painted crescents and stars on their faces, and, in Queen Anne's reign, the Marquis de Zenobia attended a party decked with sixteen patches, one a tree with lovebirds.

Like fans, these little marks developed a code. One on the forehead showed majesty, on the nose sauciness, on the mid-cheek gaiety, and near the corner of the eye, passion. A patch on the lips invited a kiss. By the eighteenth century, this signal system had grown political. Whigs put patches on the right cheek, Tories on the left. The neutral placed them on both. The mole had become a bumper sticker.

A mole can be dainty, singular, attractive. No one has ever made these claims for acne. It is the classic adolescent agony. Just as one passes puberty and starts to care about appearance, pimples erupt on the face and taint it. This jest of fate is doubly ironic, since the sex hormones trigger it.

It can be a wrenching problem, causing teenagers to retreat into themselves out of shame. Overall, acne sufferers report worse emotional adjustment than victims of malignant melanoma. Yet some acne is near-universal among young people, though it usually amounts to just a pimple or two. In one study, only 10 percent of adults recalled having had acne, though surveys prove its adolescent ubiquity.

Like moles, acne is highly hereditary. When both parents have had bad cases, their children tend to suffer as well. Identical twins show near-identical acne. Acne is not contagious and not related to vitamin deficiency, though some dermatologists recommend against eating pork and dairy products. At times it seems to worsen under stress, and doctors aren't quite sure why. Failure to clean the face neither causes acne nor cures it, since the problem lies deeper.

Acne arises when some obstacle blocks the tubes leading up from the sebaceous glands in the skin. It traps a microorganism called *Propionibacterium acnes*. The *P. acnes* bacteria multiply in the oil of the gland and secrete enzymes called lipases, which make fatty acids out of the oil. The fatty acids irritate the gland and eventually make it swell, causing a pimple. The whole process takes two to three weeks.

Acne is mainly a human ill, since we have far more sebaceous glands than other animals. These tiny organs begin making and pumping out a waxy substance called sebum at puberty.
They may be waterproofing devices, and paleoanthropologist
A. M. W. Porter suggests they protect against rain, since they
mainly appear on the head and, more sparsely, the upper body.
But as he notes, their role remains unclear. Acne is a malfunc-
tion of a mysterious body part.

Physical pressure can cause acne. Traveling salesmen have
developed it on their backs from pressing constantly against
vinyl seat covers, which makes them hot and sweaty. Others
have gotten it on one side of the face from sleeping on their
hands, or on the forehead and chin from football helmets.

Cosmetics can block the sebaceous glands and cause acne,
one reason it can arise in older women. In fact, experimenters use
cold cream and cosmetics to induce acne in lab animals. Dermatologists consider *acne cosmetica* a severe and widespread problem.

What is the cure? The ever-helpful Ovid recommends collecting "Halcyon-cream" from the nests of sea birds ("They will
complain, of course") and mixing it with honey into a paste.
The Ebers Papyrus of 1500 B.C. suggests abrading the face with
a mix of alabaster and grain. Dermatology professor Ernst
Kromayer (1862–1933) urged a more painful scrubbing, with
dental burrs and rasps. These mass pimplectomies were about as
effective as crystal healing. Indeed, they worsened the symptoms
without touching the cause. Squeezing or popping pimples risks
infection and can lead to scars, since it forces rotting material
back into the flesh.

Fortunately, a real cure is not so painful, and for decades
antibiotics like tetracycline and erythromycin have reliably
scythed away *P. acnes* bacteria. Since many strains have grown
resistant, physicians now advocate drug cocktails or direct skin

treatments with medications like benzoyl peroxide. The best current remedy is probably isotretinoin, sold as Accutane or Roaccutane. Seven million patients have taken it since it appeared on the market in 1982, and one scholar calls it "the most significant therapeutic advance in acne therapy."

In the fourteenth century Isabeau of Bavaria, later queen of France, laved her face in potions made of wolf blood, boar brains, and crocodile glands, as crones chanted arcane formulas. The ingredients have changed over time, but a sense of magic still pervades the quest.

For many women, adulthood is a time of great attentiveness to the facial skin. In fact, all it really needs is gentle soap and water, moisturizer, and, at times, sunscreen. But the American cosmetics industry is unregulated and people buy in a haze of semi-promises. They usually spend too much. A recent study in *Consumer Reports* found that brands priced under $10 worked as well or better than those costing much more.

Women used moisturizers as early as 3,000 years ago, when Egyptian ladies coated their skin with scented oils after bathing, and sometimes slathered on hippopotamus fat and a mix of honey and powdered donkey teeth. Ancient Syrians rubbed their skin with camel lungs. By the 1940s scientists had invented oil-in-water emulsions, chemicals with a capoff; once the skin absorbed the water, the oil sealed it in. In the 1970s researchers added urea to the brew. Urea is hygroscopic, meaning it pulls water into the skin and holds it there. Recent moisturizers contain alpha hydroxy acids — glycolic, lactic, and a few others — which remove rough, dead cells. These acids can also reduce wrinkles, though most products contain too little for any discernible effect.

The mud pack is an ancient beauty caress. Roman women used crocodile dung as a "face pack," a true sacrifice for one's face, and modern mud packs often come from locales with medicinal reputations. Sometimes these substances are actually helpful. When people with scabies bathed in the sulfur spring at Lourdes in France, their weepy sores healed up: a miracle. Many mud baths killed parasites, and we still use precipitated sulfur and petrolatum (vaseline) to treat chiggers, head lice, and crabs.

Moreover, diatomaceous earth and various minerals can moisten the skin, so a bather with a very dry epidermis emerges less wrinkled, cracked, and scaly. Old skin often looks muddy or blotchy because of the accumulation of dead squamous cells. A mud pack can help peel them off, leaving blood vessels closer to the skin surface and making it look pinker, fresher, younger. The effect is temporary and requires endless repetition.

Bags under the eyes are a well-known everyday curse, a leaden, raccoon-eye look that suggests necrotic flesh. What causes them?

There are many triggers. One is tiredness. When we use the eyes extensively, the body pumps extra blood to them to swab up debris and they become bloodshot. A network of tiny blood vessels also surrounds the eye, close to the skin. In exhaustion, blood may flush them too and darken the tissue. Bags can simply be bloodshot skin.

Allergies are another cause. People with sinusitis or hay fever often rub their eyes, reddening the sclera and sometimes darkening the folds under the eyelids. They can also get a crinkle on the upper nose, where the cartilage begins. It's called a "salute line," since running the hand salute-like up the nose causes it. Children with allergies often show it.

A handful of other factors contribute. For instance, our faces reflect the lifelong pull of gravity, which lengthens the jowls and deepens orbital bags. In addition, some people develop fat pads under the eyes, often related to cholesterol or triglycerides. And the contour of the face worsens the problem. Shadows fall differently on the upper and lower eyelid. The upper eyelid has curved surfaces, so the shade doesn't pool. But the flat plane below the eye is a shadow catcher.

We think of wrinkles as marks of decrepitude. They are the claw lines of age, reaching up from the grave to slowly tug us in. As Baudelaire maliciously sings:

> *Angel of beauty, do you wrinkles know?*
> *Know you the fear of age, the torment vile*
> *Of reading secret horror in the smile*
> *Of eyes your eyes have loved since long ago?*

But wrinkles don't come from age. They come from the sun, which causes 90 percent of all perceived aging in the face. Chronologically old skin is remarkably smooth, though it loses some elasticity. But sunlight slowly maims the face, causing fine and coarse wrinkles, mottling, and looseness. It also thickens skin, and the Aymara Indians of 12,000-foot-high La Paz have complexions like polished cowhide. Since the sun only strikes exposed areas, faces look wrinkled and bellies don't.

The culprit is ultraviolet rays, invisible light of shorter wavelength than visible. Ultraviolet covers the wavelengths from 200–400 nanometers. (Visible light, from 400–760 nm., is virtually harmless to skin.) When wavelengths are short, more strike the skin per second. They thus have greater energy and are more dangerous. Scientists break ultraviolet down into three categories: C, B, and A. The deadliest is UVC, with narrow pulses 200–290 nm. wide. Atmospheric ozone completely sops up this menace before it hits the earth. UVB lies in the middle of the range, at 290–320 nm., and is the most hazardous ultraviolet that reaches us, though ozone screens out 90 percent of it. UVA, at 320–400 nm., is a thousand times weaker than UVB, but about 100 times more plentiful, since ozone doesn't stop it. UVB is like a sprinkling of acid in the rainfall of UVA.

Sunlight is not the sole source of wrinkles. Smoking causes them too. One 1995 study showed that smokers were two to three times more likely than nonsmokers to have moderate or severe wrinkles. The mechanism is unclear. Some chemical in smoke may cause the harm directly, since we know it damages collagen and elastin in the lungs. It may aggravate the impact of sunlight, or narrow the arterioles to the skin, or cut down on the amount of vitamin A the body absorbs.

Wrinkles have other causes. Gravity pulls the skin down over time. Habitual expressions can mold it, so worry or anger puts vertical grooves between the eyebrows and squinting mars the eyes. People who lose teeth develop wrinkles around the mouth.

If adults sleep on their sides, says dermatologist Terry Brazell, doctors can often tell which one. "You put your head on the pillow and it folds the face there," she notes. Often a line slants outward from the eyebrow up to the hairline. If the sleeper tucks his chin into his chest, the crease may run from the corner of the mouth to the ear.

Whatever the source of wrinkles, most people simply want to annihilate them. The basic treatment involves prodding the formation of new skin. As with mud packs, sloughing off the old, dead surface yields a fresh, babyish look. The pioneer wrinkle cream, Retin-A (tretinoin), causes peeling of the outer epidermis, spurring new blood vessels and collagen. It fluffs the surface. Similarly, lasers strip away the top layer and perk up the one just below.

All such treatments must be ongoing A patient who stops using Retin-A regresses, and laser therapy is a continual resurfacing. They all remove the patina of age, but we still don't know how to make a new skin every week, and we may never know.

From Flatheads to Change Alley

In 1982 scientist Malte Andersson captured 36 male widow birds, an African species with a long, black tail. He lopped off some males' tail feathers and pinned them onto other birds. He discovered that those with the artifically lengthened tails won more mates than other birds. The widow bird's tail is the ultimate ornament, one that actually determines the reproductive success of its owner.

Every day humans try a similar trick with their faces. They have plastic surgery.

Designer faces date back to the head-molding of primitive tribes. The Mangbetu of northeastern Zaire shaped skulls, as did Indians of the American West. The Flatheads earned their English name from the custom, but it also appeared among Indians of Caribbean South America, the coast and highlands of Ecuador, Peru, and northern Chile, and the cold river valleys of Patagonia.

Along the lower Columbia and northward, Indians flattened babies' heads by squeezing a padded board at an angle to the forehead. This technique led to the wide "Chinook" head. But the Nootka and Kwakiutl created a highly elongated head, sometimes called the "Koskimo" after the Indians who devised the most extreme version.

According to Spanish writer Garcilaso de la Vega (c. 1540–
c. 1616), Indians of southern Arkansas molded heads that were
"incredibly long and tapering on top," by binding them from
birth through the age of nine or ten. Natchez shaped their heads
like miters. The Indians of the southeastern United States generally flattened their heads and ridiculed whites as "longheads."
Choctaws created plane-like foreheads by placing bags of sand on
babies' brows. Sometimes they pressed in the rear as well, which
gave them pates like bricks and made their eyes bulge.

One anonymous observer, apparently describing the
Natchez, wrote that what a mother did to achieve this effect was
"almost beyond belief." She placed the baby upside-down in a
cradle. Then she pressed clay on the top and bottom of its head,
put boards on the clay, and pressed them together as hard as possible. The child wailed and turned black, and white fluid poured
from its nose and ears. The mother let it sleep this way every
night until she had reconfigured its skull.

One scholar says Nefertiti (14th cen. B.C.), wife of the
pharaoh Akhenaton, had an artificially stretched head. The
Huns likewise pulled crania up and back. They passed the cus-
tom to the Eastern Alans, whose Crimean descendants in 1600
A.D. — some 1300 years later — were still elongating heads.
Indeed, in *Émile*, Jean-Jacques Rousseau (1712–1778) claimed
French midwives were reshaping babies' skulls.

The Phantom of the Opera has no nose and in public wears a
pasteboard prosthetic, complete with mustache. Later he tells
Christine Daaé he has invented a mask that will make him look
just like others. "People will not even turn round in the streets,"
he says. But noselessness is the source of his ugliness and his evil.

There is no denying the emotional damage of an injury or deformity on the face. It is far worse than elsewhere. Unlike a withered arm or scarred leg, it directly harms self-image. It may also alter others' opinions of one. When Islamic leaders wanted to stop people from worshiping the Sphinx in the fifteenth century, they cut off its nose. It worked.

But the human face, as it happens, is well-suited for reconstruction, since it has many veins and arteries. The rich supply of blood hastens recovery, though it often causes postoperative bleeding, even from simple exercise.

In the West, plastic surgery dates at least to the thirteenth century, when Lanfranco (d. 1315) and Theodoric, Bishop of Cervia (1205–1298) described nose-enhancing operations. The modern pioneer was Gaspare Tagliacozzi (1545–1599), author of the classic *De Chirurgia Curtorum* (1597). Tagliacozzi himself was a popular surgeon, and his method for rebuilding noses lost in a duel might have helped the Phantom. It involves bringing down a flap of forehead skin, and surgeons still use it. The Church frowned on him for improving God's work, and after his death priests unearthed his body and reburied it in unhallowed ground.

Among the many triumphs of plastic surgery are the closure of cleft palate (usually performed in babyhood, to preclude trauma), the rebuilding of the saddle-nose from congenital syphilis, and the restitution of faces injured in war or accident. In addition, it can normalize less serious, but still psychologically debilitating features, like receding chins, bump noses, birthmarks, port-wine stains, and prominent ears.

Old wives' tales held that ears perpendicular to the head arose from letting babies sleep on their sides, with their pinnas bent forward. In fact, they stem from a large concha or a small anti-helix, both of which thrust the ear flaps out into view. In March 1881, Edward Talbot Ely (1850–1885) performed the first modern operation for reducing them. Parents had brought in a twelve-year-old boy who complained of teasing because of his prominent ears. Ely made a cut where the ear meets the skull and removed some cartilage. The ear now stood further back. Surgeons have since refined the procedure, and it is a staple of the field.

A blepharoplasty is a lid job. In the late twenties or thirties the upper eyelids begin to sag and "crow's feet" appear at the corners. In older persons the eyelid skin loosens further and can droop over the outer eye. A lid job can clean up these problems, making the eyes seem larger and their owners younger.

In whites and blacks, a furrow called the superior palpebral fold runs across the eyelid. About half of Asians have this fold too. Some who don't have been getting the lid job called a *bookoochai*, after the Singapore surgeon Boo Koo Chai who invented it in the 1960s. It makes the eye seem both larger and more Western. The surgeon cuts an incision in the lid where the fold will go, removes fat, stretches the skin, and reshapes the muscles. The operation takes about an hour per lid and can cost $3,500. Most clients are women. Sometimes Korean-American parents bestow this gift on their daughters on graduation from high school or college.

The procedure is controversial. Some Asian-Americans claim that women who undergo it are trying to imitate white features rather than accepting their own. Ronald Matsunaga, a Beverly Hills plastic surgeon who specializes in these operations, denies the charge. His clients, he notes, simply seek bigger and more alluring eyes. He points out that the fold isn't Western, since many Asians are born with it, and that patients keep their other Asian features: the high cheekbones, wide face, flat nose.

They also retain the epicanthus web, the fold of skin that covers the eye corners and exists in about 90 percent of Asians. The sides in this debate are very far from convincing each other.

Da Vinci felt the nose determined the basic impression a face makes, and Schopenhauer said, "The fate of so many women depends on a slight up or down curve of their nose." Hence the rhinoplasty, or nose job. Thomas Pynchon devoted a chapter of *V.* to a clinically detailed description of it. Michael Jackson, Tom Jones, and Phyllis Diller have had nose jobs, and indeed it explains the paucity of long noses in the Western world today.

Rhinoplasty dates back at least to the extravagant and boastful Johann Friedrich Dieffenbach (1794–1847), but Berlin surgeon Jacques Joseph (1865–1934) generally receives credit for developing the modern operation, which he first performed in 1896. Joseph worked in this field when his peers deemed it trivial, and he actually received a suspension from his university post because of it. Like other pioneers in the area, he did not suffer from a weak ego. A visiting American surgeon once asked him to converse in English. "No," he replied. "It is beneath my dignity."

A nose job can take place almost entirely inside the nostrils, leaving no scars. Cartilage makes up much of the nose, and surgeons can remove it much more easily than bone. Some operations are fairly simple. To eliminate the hump in a nose, physicians painlessly cut out cartilage from within the nose. They can shorten a nose by removing the flesh from the tip, and narrow one by taking cartilage from the sides of the wings. They emphasize that they can't always create a perfect nose, only an improved one.

Often surgeons add implants to swell parts of the face. When a nose is too small, as with the saddle-nose, a small plug can gracefully boost it. Implants can help correct a receding chin, and Marilyn Monroe had such an operation. They can also enhance cheekbones. Surgeons have used ivory, human bone, sheep cartilage, peanut oil, and polythene as implants, and occasionally bone or cartilage from the patient, but silicone rubber has become the material of choice.

Patients happily place their double chins on the sacrificial altar of beauty. In a chin-lift, surgeons pierce the flesh and remove fat, commonly with liposuction. The brow-lift, a similar procedure, makes the eyebrow more arched and youthful-looking. Surgeons can also alter mouth size and reduce the prominence of lips. Recent advances in laser surgery let physicians remove bags under both eyes in fifteen minutes, with just local anesthesia.

The face-lift is probably the most glamorized operation in this field. It began with the efforts of the American Charles Conrad Miller (1880–1950), who was both a visionary and a quack. Miller made some advances, but he also tried to eliminate wrinkles with injections of paraffin and even a mix of silk, celluloid, floss, and gutta-percha which he made in a vegetable grinder.

The modern face-lift is a little like tugging a loose mask tighter onto the face. The surgeon makes a long incision behind the ears and pulls the facial skin up and back, stretching it over the underlying structure. A face-lift is most effective at removing the sagging skin around the jowls and upper neck. It usually doesn't affect fine wrinkles, crow's feet, or the little vertical lines near the mouth. It also won't eliminate double chins or the twin bands like shelf brackets that can appear in the aging neck. Best performed in one's forties as preventive maintenance, a face-lift normally lasts five to ten years.

Not every customer has rejoiced over the results of plastic surgery. When a hair stylist botches a cut, the hair grows back, but face surgery is sculpture. Yet even when the job is perfect, psychological difficulties can arise. Some people blame their problems on facial flaws and see plastic surgery as a royal road to happiness. Often they are elated as they leave the hospital, but grow depressed as time passes and they realize that their lives haven't changed much. Moreover, we identify with our face. After surgery, people may feel their face "doesn't look right" or "doesn't belong to me." The key, physicians say, is self-acceptance beforehand. A beautiful face won't bestow it.

In John Varley's novel *Steel Beach*, which takes place on the moon several centuries hence, identity has become a toy and beauty an easy commodity. People live to be over two hundred, yet unlike the struldbruggs they remain alert and youthful. They routinely change sex. They drink potions that alter eye color and hair structure overnight as they sleep. And they have designer bodies and faces, genetic resculptings to order. They can choose their looks.

At one point protagonist Hildy Johnson goes down Change Alley to find a face designer. He has decided to reconfigure his nose and ears, and slightly increase the fullness of his lips. It scarcely requires surgery. He simply quaffs a saline potion filled with trillions of nanobots. Nanobots, or nano-robots, are molecule-sized smart machines. Once they have a snippet of one's genes, the designer injects them into the bloodstream and they find their way to the right destination. Some enter the facial DNA itself and rearrange it. The larger ones have tiny motors, scrapers, and arms, and they excavate or add to organs like the nose, using materials hauled in by other nanobots. When finished, they disengage, drain from the body, and return to the bottle for the next customer.

It actually seems feasible, though certainly not tomorrow. Change Alley would end discrimination by looks, remake the cosmetics industry, and fill the world with living masks. What kind of faces would we find? Would we see flatheads and miterheads, small eyes and large, or would faces converge in a single model?

A Thousand Ideals?

Poet Abraham Cowley (1618–1667) called beauty "thou wild fantastic ape/ Who dost in every country change thy shape!" And indeed standards of beauty seem to vary remarkably across cultures.

The oval face, for instance, is the peak of perfection in Greek and Roman sculpture, and it dominates much of the painting of the Italian Renaissance. Praxiteles' Aphrodite of Cnidus has an oval face, as do most heads by Raphael and Correggio. In the days of black-and-white movies, cinematographers likewise exalted the oval face, partly because it was easier to light and it shot symmetrically from all angles.

But in the classic Mideast, connoisseurs esteemed the round face. In "Julnar the Mermaid and Her Son Badar Basim" of *The Thousand and One Nights*, King Shahriman marvels at the face of his pretty new concubine, which has "the rondure of the full moon or the radiant sun shining on a clear day." Throughout this vast work, a beautiful face is the moon or a pearl, a circle of soft glow.

Among the Ashanti of Africa, pregnant women carried *akua mma*, special dolls to ward off deformity in the child. The dolls embodied the Ashanti ideal of beauty: round flat face, high forehead, small mouth, long neck. Mothers-to-be gazed intently at these faces, hoping such scrutiny would improve the child's lineaments.

One might think all peoples prize brilliant teeth, and certainly dentists today whiten plenty of teeth with hydrogen peroxide and laser beams. Yet Natchez and some Choctaw Indians blackened their teeth with ashes and herbs to beautify them, and a few South American tribes did the same. Numerous peoples file or chip their teeth, and David Livingstone reported a tribe in Africa that knocked out the two upper incisors, for beauty.

For centuries people generally felt society shaped ideals of beauty, and Darwin seemed to agree. Yet explorer/linguist Richard Burton (1821–1890), a man with firsthand experience of cultures around the globe, believed a single standard was universal. In the past ten to fifteen years, scientists have probed the issue, usually asking individuals from different cultures to rate the same photos for beauty.

The results have been consistent: People everywhere have virtually the same sense of facial beauty. Americans, Koreans, and Chinese living in the United States have agreed about the esthetic merits of people in their own groups and in others. Chinese, English, and Indian women preferred the same Greek men. Blacks, whites, and Chinese have concurred on white and Chinese men and women, and newly arrived native Asian and Hispanic students have agreed with white Americans on the attractiveness of Asian, Hispanic, black, and white women.

In the latter study, researchers considered the possibility that Western media had spread one ideal of beauty across the globe. Hence, they tested Taiwanese with little exposure to Western culture. These subjects agreed with the rest. They also tested black Americans for body preference. Black men favored a heavier figure, a higher ideal weight, and larger buttocks than white men. But if the media shapes facial preference, researchers noted, it should shape body preference too.

In fact, the response to beauty is the same across almost every other divide as well. Britons in various occupations and walks of life agreed on it. Males can recognize a handsome man, and women a pretty female. People of every personality type agree. Attractive raters concur with homely ones, though in one study the extremely good- and bad-looking gave everyone lower scores. The rater's age doesn't matter either. Children of seven, twelve, and seventeen agreed with adults, and preschoolers rated other preschoolers as adults did. Indeed, the skill is there in nonage. Both white and black infants looked longer at attractive men and women of either race. It is hard to see how babies could have learned to make this discrimination in their first few months of life. As several researchers have noted, this fact argues for an inborn template for beauty. It's in the genes.

Indeed, it would be amazing if it weren't. As evolutionary psychologists note, beauty affects gene transmission too critically to become the plaything of culture. One might as well let a monkey drive a limousine. Beauty is a governor to keep the species on course, and if culture had determined it in any of our hominid ancestors, they would have thinned out or vanished. We have an inborn ideal of beauty because it has somehow helped us survive.

Then why do cultures seem to have different standards? One reason is that we can confuse beauty with prestige. Status boosts ratings of attractiveness, and status can attach to almost anything. Canadian Indian women deemed their labrets appealing. Where only the wealthy grow obese, fat may become prestigious and desirable.

Fashion tweaks notions of beauty. In the fifties, women considered beehive hairdos stylish, hence alluring. They would not have worn them otherwise. Tattoos and body rings are in vogue today and may prove equally ephemeral. Artistic conventions can last much longer. Butterfly effects can establish them and they can linger on, recipients of lip service and unrelated to popular opinion. And custom can sanctify even drastic measures like head-shaping, so deviations feel abnormal.

Moreover, scientific paradigms long hindered research. One spinoff of cultural determinism was the eye-of-the-beholder idea that beauty could take any form. It discouraged serious experiments for decades until genetics began to force its shambling anabasis in the 1960s. As one scientist said in 1964, "Except for some arbitrary beauty-contest conventions about ideally female dimensions, we know less about attractive stimuli for men than we do about those for fish."

Moral passion has contributed. Cultural determinism lingers on, and some feminists still assail beauty as a male invention. Naomi Wolf, in her best-selling *The Beauty Myth* (1991), dismisses all evolutionary arguments without articulating or responding to them, and calls beauty a currency standard. Like any currency, it is thus a tool of politics. The notion has appeal, since beauty is unrelated to so many other merits and seems to improve or diminish lives unfairly. If it is a cultural artifact, people can modify and transcend it.

Finally, false reporting has afflicted the field. Travelers have always loved colorful stories and commonly exoticize residents of distant lands. Early visitors to Patagonia, for instance, solemnly reported a race of giants. Other voyagers have returned with tales of bizarre beauty standards. Darwin cites Hearne's 1796 remark that Northern Indian men prize a female with "a broad, flat face, small eyes . . . a low forehead, a large broad chin, a clumsy hook nose, a tawny hide, and breasts hanging down to the belt." Later researchers found absolutely no support for this claim among any tribe in the region, and indeed these people judge looks much like everyone else. Beauty is genetic, global, and everywhere recognized.

A Frog's Idea of Beauty

In Gabriel García Márquez's *One Hundred Years of Solitude* (1968), Remedios the Beauty inspires whirlwinds of longing all about, yet seems oblivious to them and immune to love. Her beauty is meaningless, a lure to the void. And indeed, good looks in general seem meaningless, a kind of magnet to whatever surprise lies behind. As Wilde says of beauty, "The true mystery of the world is the visible, not the invisible."

Why does it matter? How does beauty keep us fit evolutionarily?

First, it is species-oriented. Voltaire said, "Ask the frog what is beauty . . . ; he will answer that it is a female with two great round eyes coming out of her little head, a large flat mouth, a yellow belly and a brown back." The astute Cotta in Cicero's *The Nature of the Gods* (44 B.C.) says, "Do you imagine that an eagle, or a lion, or a dolphin prefers any other shape to his own?" And Firenzuola observed, "For a camel a hump is a thing of grace, for a woman, a misfortune. This can come only from a mysterious order in nature that, in my opinion, the human intellect cannot fathom."

But the human intellect has.

Consider forest guenons, small monkeys who dwell in the upper branches of trees. Five closely related species live beside each other, and each has a different countenance. Brilliant eye-rings, blobs, and streaks of color adorn their faces. These distinctions help an animal recognize its own kind and prevent hybridization, which can lead to sterile offspring. The device works. These guenons intermix very little.

Evolution has selected for these facial patterns, and it doesn't matter what they are, exactly. Guenons feel attracted to the faces of their own species and not those of others. And if a guenon wondered why one pattern should seem "beautiful" and not another, the question might seem baffling too.

In all species it is critical to recognize potential mates. The signals vary widely. Birdsong not only tells one bird that another of the opposite sex is nearby, but attracts the bird. Variations in the caliber of song may be tantamount to beauty or ugliness. Dogs in heat send out chemical messages. Male fireflies signal with a precise pattern of light flashes, varying in duration, frequency, and color, and females inherently know which pattern to respond to. Antelopes recognize potential mates by their antlers, which come in many sizes and shapes.

In primates, the face is the natural place for such signals, since they pay such attention to it. And their faces vary markedly. De Brazza's monkey (*Cercopithecus neglectus*) has an orange "diadem" and a white beard. The mandrill (*Mandrillus sphinx*) has a vivid scarlet nose and brilliant blue swellings on either side of it, a striking color pattern repeated on the genitalia. The red uakari's face looks like an all-day sunburn. The white-faced saki has a nimbus of white hair around a black rounded triangle containing its nose and mouth.

Why would people need such a system? No species like us exists on earth, and no confusion seems possible. But many different hominids have walked the earth in the past. Our lineage is not a line, but a bush, with many truncated limbs, and our ancestors may well have used the face to distinguish themselves from others. Indeed, *Homo sapiens* lived beside the Neanderthals for thousands of years, and we differed from them mainly in the face.

The role of the face in recognizing mates partly explains why facial deformities can seem so devastating. The lack of a chin, for instance, affects us viscerally, and differently from lack of even a leg. We feel the individual is somehow maimed in his very personhood.

If beauty flags a prototypical *Homo sapiens* mate, then averageness should be a key feature of it. The most ordinary faces should seem the most extraordinary.

The notion of averaged beauty goes back to antiquity. According to legend, the Greek artist Zeuxis (fl. 400 B.C.) fused several attractive models into his celebrated painting of Helen, making this perfect beauty a composite. But scientifically it began with one A. L. Austin, who in 1877 wrote Darwin that two portraits placed in a stereoscope and merged into one produce "in every instance a decided improvement in beauty." In 1878 eugenics prophet Francis Galton (1822–1911) attempted to create a prototype criminal face. He superimposed many malefactors' faces atop one another on a photographic plate, making composite portraits, and he too noticed the effect. It thwarted him. The average criminal seemed too good-looking.

In 1990, a team of experimenters fed the faces of 96 males and 96 females into a computer, which merged them into averages based on 4, 8, 16, and 32 subjects. These faces tended to look androgynous, symmetrical, and somewhat vague and blurred. Then 65 students judged their attractiveness. They almost always rated the 16- and 32-face composites as more appealing than the individual faces.

Mate recognition partly explains this phenomenon, but the researchers observed that other factors could also contribute. For instance, an average face is one without abnormalities, which often mean defects. Moreover, they noted, it could indicate a balanced makeup of genes, which might better protect against illness.

Yet averageness is not enough. Indeed, while merged faces are attractive, they aren't the acme. In one 1991 study, researchers examined the faces of international beauty queens and found them better-looking than a face averaged from the population. In another 1994 study, investigators blended 60 female faces into one on a computer. Next they fused the 15 most attractive into a second face. The overall-composite and the attractive-composite differed, and subjects found the latter more alluring. Researchers identified the differences and exaggerated them to make a third face. Subjects found this caricature better-looking than either. The researchers replicated these findings with Japanese women's faces.

How did the attractive faces differ from the average? It has to do with the reason woman have always needed a beautiful face more than men.

Beauty and the Clock

In November 1972, a photo of a beautiful Swedish woman named Lena Sjööblom appeared in *Playboy*. She peers over her shoulder at the viewer with deep liquid eyes and the hint of a flirtatious smile. Scientists at the University of Southern California digitized this face, and Lena became a standard for tests in image processing, known and used by engineers around the globe. Most of them are, of course, male. It's hard to imagine a "Len" becoming a similar standard in any female-dominated profession.

Research confirms what most people know anyway: A woman's looks matter more than a man's. A 1966 survey of a thousand college students asked, "To what extent is it important that your date be good-looking or attractive?" Among the men, 22 percent checked "very important"; among the women, 7 percent. The women more often sought qualities like status and high grades. In personal ads, women are more likely to mention that they are attractive, and men that they seek attractiveness. A 1951 survey of 190 cultures found the same pattern everywhere. Men cared more about looks, women about prowess.

Literature teems with tales of ugly men whose façades good women or society sees through. The Beauty comes to love the Beast. The Phantom kisses Christine and, poignantly, she doesn't shy away. Quasimodo's face repels more people than his hump, and he is highly sympathetic. Quoyle gains a little peace and acceptance in E. Annie Proulx's *The Shipping News*. And movies like *The Elephant Man* and *The Man Without a Face* teach that beauty lies in character, and both have "man" in the title.

Such tales of moral triumph rarely have a woman as their focus, especially when males write them. One exception is "Brigitta," by Austrian writer Adalbert Stifter (1805–1868). Brigitta grows up ugly, but slender and strong, and the handsome, discerning Stephen Murai falls in love with her. They later separate and she moves out to the Hungarian steppe, which she makes bloom. Brigitta's beauty lies within, emblemized by her talent for bringing treasures out of a wasteland.

Women writers have treated this theme more subtly and perceptively. Their sympathetic heroines are often plain, rarely deformed. Pretty women flutter about nearby, and they tend to be superficial, conceited, even mean. Georgina Reed in *Jane Eyre* is nice but shallow. Rosalie Murray in *Agnes Grey* is atwitter with petty cruelty. They can lure men, but often to their tragedy. As George Eliot says in *Adam Bede* about the babyfaced Hetty, "People who love downy peaches are apt not to think of the stone, and sometimes jar their teeth terribly against it."

Why the skew? As evolutionary psychologists note, it's a matter of biology, a question of which mates leave behind the most DNA, for theirs has formed our faces and longings.

And men attracted to nubile women have done better than vice versa. One reason is that women's fertility drops faster than men's over a lifetime. Indeed, the female rate is predictable across the globe. By ages thirty to thirty-four, women are 85 percent as fertile as they were between twenty and twenty-four, and the rate declines to 35 percent by forty to forty-four, and 0 percent after fifty. Among men, it tapers off more gently, reaching 90 percent by forty-five to fifty and about 80 percent after fifty-five. So males who mated with older women passed on less coding, while females could mate with older men without the same problem.

Youth also implies health. Thus young women are more likely to survive childbirth and — critically — to live to raise the child. They are also slightly less prone to bear children with birth defects. Our species evolved on a savanna where medicine was folkloric, sepsis a mystery, and parasites even deadlier than today. Hominid health was a fragile matter, and an evolutionary sieve.

The genes that crossed the generations tended to come from young women who not only were healthy, but who signaled health with traits like a smooth face, lustrous hair, sparkling eyes. Indeed, cues of youth and health are almost universal desiderata in human beauty, and in *The Mayor of Casterbridge* (1886), Elizabeth-Jane Henchard possesses "that ephemeral precious essence, youth, which is itself beauty.' Old faces sag and wrinkle, and fat drowns facial contours. We rarely seek these features in mates (though we accept and even love them), and this human interest in youth is unique among animals.

In one study, males judged women as less attractive with age, but the female response to older men was mixed. Health and looks matter in men, of course, but youth is less important, since powerful males are often older. They can protect a child better. Among women, genes have prevailed favoring less the appearance of men than their ability and willingness to provide for offspring. Hence women care more about a man's status and wealth, his height and physique, and his commitment to her. Here is Jane

Eyre on Rochester: "My master's colourless, olive face, square, massive brow, broad and jetty eyebrows, deep eyes, strong features, firm, grim mouth—all energy, decision, will—were not beautiful, according to rule; but they were more than beautiful to me." Yet he is rich and powerful, and he radiates authority.

What are the features of a good-looking face? In the 1994 study of averages, the averaged beauties had higher cheekbones, thinner jaws, and larger eyes relative to face size than the averaged faces. The distances between chin and mouth, and mouth and nose, were also shorter. That is, their lower faces were smaller. They were more childlike.

Other studies have, in general, replicated this finding. Michael Cunningham conducted one of the best known in 1986, in which male undergraduates rated full-face photos of females. He found that the ingredients of a gorgeous female face were:

> wide cheekbones
> narrow cheeks
> broad smile
> wide nostrils
> wide eyes
> high-placed eyes
> eyes set far apart
> large pupils
> high eyebrows
> small nose

The first two, wide cheekbones and narrow cheeks, are signals of sexual maturity, like full breasts, and Cunningham suggests they may discourage pedophilia. The third was a broad smile. A smile, of course, is an expressive feature, but it can range in size from a small crescent to a toothsome dazzle that fills the lower face. Wide nostrils are smile indices, since a grin expands them.

The rest of the list catalogued babyish features, all but two vulnerable to age. Young eyelids are firm and make the eyes look large and round, but the years loosen this skin and narrow the eye. Women have bigger pupils than men, and in both sexes the pupils slowly shrink after adolescence. The young have arched eyebrows, which sag toward the eye over time. Most infants and children have turned-up noses, but with age the tip of the nose slowly droops, as if melted.

Once males developed a genetic preference for a young female face, women had an interest in feigning youth. Cosmetics and plastic surgery conjure the illusion today, but our DNA has also checked in. Indeed, the babyface is probably a genetic trick. By falsely signaling youth, it attracted more powerful men and left behind more copies of its genes. Hence faces, and women's in particular, evolved to look unusually young.

Another study approached this matter from a completely different angle.

Victor Johnson and Melissa Franklin of New Mexico State University wanted to find out which faces would emerge as ideal beauties using evolution, the process that created them in the first place. They devised an experiment in which subjects "evolved" their own best-looking faces on a computer, using a genetic algorithm.*

---

*A genetic algorithm is software that re-creates natural selection. Starting from an initial population, it chooses some individuals for mating and shunts others aside. It breeds a new group, selects further individuals for mating, and continues down a series of generations. Thus the population evolves. Normally the software chooses the breeders, but in this case the subjects did. With each mating, this program varied the offspring with crossover and mutations.

They began with 30 female faces chosen at random. The subjects, 20 men and 20 women, ranked the appeal of each face on a scale of 1 to 10. In this artificial environment, beauty was fitness, period. Researchers mated the most attractive face with one of the other 29. In each new generation, if any face rated higher than the lowest-ranked face of the old group, it supplanted that face. Hence the population remained at 30 over the generations.

The researchers cycled through the generations until a subject had either rated 100 faces or deemed one a perfect 10. Most subjects evolved a 10 in around 80 minutes, after seeing an average of 77 faces.

These computer-bred beauties looked much like the pretty women in other studies. They had smaller mouths, yet fuller upper and lower lips. They had daintier chins. Eye-to-chin distance was shorter, as were both eye-to-mouth and mouth-to-chin distances. This short lower face, the investigators said, resembled that of a young girl of perhaps eleven. Female subjects evolved faces almost identical to male, though they gave them a larger lower lip.

The researchers noted that the beauties had traits reflecting both high estrogen levels, like fuller lips, and low androgen, like smaller chins. This combination, they suggested, signaled high fertility. Most women in hominid tribes were likely to find mates, they argued, so the key to swamping future generations with one's genes was fertility. Beauty means bounty in offspring.

Do these arguments seem to fully explain the importance of facial beauty? If not, consider a notion the great statistician Ronald Fisher (1890–1962) suggested many decades ago. He dubbed it the sexy-son hypothesis.

The sexy-son hypothesis is a momentum effect. Alluring factors tend to stay alluring, Fisher said, since conformism here makes good evolutionary sense. If you are a female, and all other females favor an average face in a mate, then you should too, since your children are more likely to have average faces. They'll be more in demand, and you'll have more grandchildren. It almost doesn't matter what the factor is. The reasoning is the same. Hence a peahen should choose the peacock with the most spectacular tail, because he's in demand and his children will be too. The peahen who prefers a mate with a drab tail may have a rich relationship and many offspring, but her genes could dead-end in that next generation. Anyone will pass on more coding if he or she mates with someone likely to have attractive children.

Beauty is not a single quality. It involves a mix of signals, especially of averageness, sexual maturity, youth, and health.

And, as everyone knows, that's just the surface.

The Deep Aurora

"If Cleopatra's nose had been shorter," Pascal said, deeming short noses ugly, "the whole face of the earth would have been changed." But Cleopatra (69–30 B.C.) was not a beauty, at least according to her portrait on coins, which we might expect to flatter her. She has a large mouth and a hooked nose, which seems to grow longer on coins issued later in her reign.

Plutarch (46?–120? A.D.), who lived a century later, agreed, but added that "the charm of her presence was irresistible, and there was an attractiveness in her person and talk, together with a peculiar force of character which pervaded her every word and action, and laid all who associated with her under its spell. It was a delight merely to hear the sound of her voice."

Louise de La Vallière (1644–1710) did not have an especially good-looking face, by all accounts, but her sweetness and radiance won the love of Louis XIV, a man who had his choice of courtly beauties. She bore him several children, and when his attentions finally turned elsewhere, spent the last thirty-six years of her life in a nunnery.

George Sand (Aurore Dupin, 1804–1876) once wrote, "I had but one instant of freshness and never one of beauty," and indeed portraits reveal her face as somewhat homely. But she was brilliant, warm, and charming, and she had famous love affairs with poet Alfred de Musset and composer Frédéric Chopin, along with at least twenty other men, such as Prosper Mérimée. De Musset, an experienced gallant, declared her the most feminine lady he'd ever met.

Even Theodora, famed for her beauty, did not become empress by looks alone.

Every face is alive with the self behind it. Faces can't help expressing the mind, and when we say we love a face, we mean the soul that animates it. Theodora's facial contours only gave her an opportunity. Cleopatra's didn't stand in her way.

"I cannot say often enough how much I consider beauty a powerful and advantageous quality," noted Montaigne in a carefully worded passage. "We have no quality that surpasses it in credit. It holds first place in human relations; it presents itself before the rest, seduces and prepossesses our judgment with great authority and a wondrous impression."

It presents itself before the rest. Indeed, most of the studies on beauty noted earlier examined people's reactions to it after brief initial exposures, where it has peak impact. But beauty is a curious phenomenon, one of permeable, shifting boundaries. It imbues its possessors with the golden light of competence and likability, but competence and likability also kindle beauty. Self can alter contour.

In one study, experimenters wondered how judgments of physical attractiveness would change as strangers grew acquainted with each other. They brought people together in one-hour meetings over four days. On the first day, researchers attributed scores on beauty 32 percent to the facial contours seen and 22 percent to the perceiver.* The raters were somewhat objective. On the second day, the balance flipflopped. Now the scores came 23 percent from the source and 33 percent from the beholder. On the third day, the ratio was 26/34, and on the fourth and final day, 23/48. The appraisals grew increasingly subjective and raters more and more disagreed among themselves. Indeed, we all know how people we like seem better looking to us, as if our minds rearrange their faces on the sly.

In romance, looks matter most to those who see male and female roles more traditionally, who agree with statements like "It's all right that most women are more interested in getting married than in making something of themselves" and "A wife shouldn't contradict her husband in public." In fact, one 1981 study found that liberal men not only did not rate attractive women as more competent, but actually gave a slight edge to homely women.

Yet even among traditionalists, no one loves just a face. In 1993 David Lykken and Auke Tellegen of the University of Minnesota asked the fiancés of identical twins to rate the other twin. Though 39 percent said they liked the other twin, 30 percent said they didn't. Only 10 percent said they "could have fallen for" the other twin — 13 percent of the husbands-to-be and 7 percent of the wives-to-be. This despite the clonal near-identity of such twins.

---

*The missing percentage reflects both favoritism and error, which the researchers couldn't disentangle.

Beauty is a deep aurora. It makes genetic sense. If the genes restricted our sense of beauty to facial contour, we would consider a crass, selfish, grating person just as attractive as his kind and thoughtful twin. Yet such a person would make a poor parent, raise fewer children, and reduce the spread of pure-beauty genes. At the same time, genes which placed a premium on kindness would likely propagate. A person who is kind to others will likely be loving and considerate to offspring. And indeed a majority of both sexes rate kindness as more important than looks in a mate.

Thus love, and feeling loved, can create beauty out of nothing. When Rochester half-jokingly asks Jane Eyre to give him a potion that will make him a handsome man, she responds, "'It would be past the power of magic, sir'; and in thought, I added, 'A loving eye is all the charm needed.'" The morning after she accepts his proposal, she herself looks in the mirror and feels her face "was no longer plain: there was hope in its aspect and life; and my eyes seemed as if they had beheld the fount of fruition, and borrowed beams from the lustrous ripple." A frog kisses a frog, and they become prince and princess.

Other factors abet looks. As we've seen, prestige is one. Contemporary writers praised Elizabeth I as beautiful, a term they might have withheld if she were a scullery maid. In one 1996 study, people ranked individuals from high-status countries like Japan as more attractive than those from low-status nations, and also believed good-looking people were more likely to come from prestigious countries.

Hence there are no ugly goddesses, and indeed legend is the best cosmetic. Marlowe's Helen had a "face that launch'd a thousand ships/ And burnt the topless towers of Ilium." But was she physically beautiful? We don't know. Homer wrote the *Iliad* 400 years after the event, and she may have looked like Cleopatra. Myth would have beautified her anyway. Even the monstrous Medusa grew alluring over time. Her transition to maiden began as early as the fifth century B.C. and by Roman times she was gorgeous.

Sexual arousal increases ratings of looks, by both men and women. So it seems does desire. In one 1979 study, researchers wondered how perceptions of beauty would change over an evening in a bar. So they approached patrons at different times of night and asked them to rate others' attractiveness. They discovered that people of the opposite sex drew higher ratings as closing time neared. And, intriguingly, there was no correlation with alcohol consumption. But people rated members of their own sex — the competition — very accurately.

And individual tastes vary, despite the genetic template. Longfellow said, curiously, "I dislike an eye that twinkles like a star. Those only are beautiful which, like the planets, have a steady, lambent light, are luminous, but not sparkling." In a 1971 study, researchers showed 207 subjects pictures of faces in groups of six. There was no face that someone didn't choose as the most attractive of its group. Researchers, aware of this subjectivity, don't rely on their own opinions of looks when they do experiments, but rather poll people and use a consensus.

Voice, intelligence, and familiarity all affect beauty. We see the self more plainly on video and, intriguingly, subjects not only rate faces on video as more beautiful than those in photos, but they disagree about them more. Indeed, the very expressiveness of the face — its merriment, disappointment, trembling hope — affect our judgment of its beauty, and we find expressive faces more alluring. Scientists are careful to use the term "physically attractive" in this research, because people are "attractive" in all kinds of ways.

Beauty goes before the rest. Feminists worry about its role, and indeed our television culture has grossly exaggerated its importance. Television shows surfaces, often briefly. We see a model in a TV ad for fifteen seconds, so she remains an utter stranger. It has thus intensified demand for contour virtues like physical beauty. At the same time, we see more beautiful faces than ever before, and unfortunately some people compare themselves to these professionals. They are less satisfied with themselves and feel they have more to live up to.

But beauty is not just a matter of face, or voice, body, or physical grace. People are beautiful because of their character, their insight, their ability to delight, their capacity for affection. A nonphotogenic person can seem very beautiful, because personality imparts a shimmer, and a photogenic person can seem plain.

Faced with such results, some scientists have suggested that, in the course of evolution, beautiful-is-good has become inextricably bound with good-is-beautiful, so that they mutually reinforce each other. But Sappho said it much earlier: "What is beautiful is good, and who is good will soon be beautiful."

Notes

Genesis

4   *Babies just nine minutes old, who have never glimpsed a human countenance, prefer a face pattern to a blank or scrambled one.*

In this arresting study researchers tested forty newborns. "All persons with whom the infant could have had visual contact were capped, gowned, and masked," they wrote. "The only potentially significant visual experience would have been fleeting exposure to the eyes of the experimenters or delivery room personnel." They showed these infants: 1) a schematic face with normal features, 2) two schematic faces with jumbled features, and 3) a blank head. The babies turned their heads and eyes significantly more often to follow the normal face. The team concluded that babies come into the world "predisposed to respond to any face," because the trait was evolutionarily adaptive. Carolyn C. Goren, Merrill Sarty, and Paul Y. K. Wu, "Visual Following and Pattern Discrimination of Face-Like Stimuli by Newborn Infants," *Pediatrics*, vol. 56, no. 4, pp. 544–549, 1975, p. 547.

A 1991 study replicated these findings in infants 37 minutes old. Mark H. Johnson, Suzanne Dziurawiec, Hadyn Ellis, and John Morton, "Newborns' Preferential Tracking of Face-Like Stimuli and its Subsequent Decline," *Cognition*, vol. 40, pp. 1–19, 1991.

4   *Similarly, monkeys raised in utter isolation can identify photos of their own kind.*

In 1966 primatologist Gene P. Sackett reared eight rhesus monkeys in isolation, and beginning with their 14th day, he presented them with slides of other rhesus monkeys, as well as of controls. Pictures of monkeys elicited much greater response than those of controls. Moreover, at age 60 to 80 days, the monkeys began reacting to photos of threatening adult monkeys with fear, withdrawal, and rocking. Their trepidation peaked around 90 days and declined by 110. Sackett suggested that monkeys have "innate recognition mechanisms" for faces and expressions. "Monkeys Reared in Isolation with Pictures as Visual Input: Evidence for an Innate Releasing Mechanism," *Science*, vol. 154, pp. 1468–1473, December 16, 1966.

13 *When an animal swims regularly in one direction, the head becomes its leading edge.*

In contrast, consider the squid. It can dart forward, backward, or sideways, since it moves by jet propulsion, squirting water out of a nozzle it shifts about. Since it has no fixed leading edge, its eyes lie midway down its length.

13 *A forward mouth swallows food easily, through simple momentum; a mouth astern would recede from it.*

Weapons like teeth, saws, swords, horns, and spikes also appear up front, where first contact usually occurs, momentum is greatest, and the gullet lies close.

14 *The taste buds lie within the mouth,*

Catfish and other bottom dwellers who swim in murk can also taste food on their skin, fins, and "whiskers."

15 *We don't have eyes in the back of our heads because by turning our necks and eyes we can see the whole panorama anyway.*

That is, their cost outweighs their benefits. Of course, it's possible occipital eyes have simply never arisen, so nature hasn't conducted the test.

Some creatures possess wraparound sight. The dragonfly has enormous eyes like aviator goggles, which contain up to 30,000 lenses. It can see almost everywhere at once, and snaring one can be a feat.

15 *A long, liana nose is bizarre, yet the elephant grew one because its head can't reach the ground.*

Asian elephants touch their sensitive trunk tips to earth to detect vibrations from running feet or hooves. They can also seize morsels as small as a grain of rice. L. E. L. Rasmussen and Bryce L. Munger, "The Sensorineural Specializations of the Trunk Tip (Finger) of the Asian Elephant, *Elephas maximus*," *Anatomical Record*, vol. 246, pp. 127–134, 1996, p. 128.

15 *The great flanged face of the hammerhead shark is one of the oddest in the vertebrate world. It may have arisen to spread the nostrils farther apart, increasing the difference between odor levels in each, so these sharks could better track the origin of delectable scents.*

See Peter B. Moyle and Joseph J. Cech, Jr., *Fishes: An Introduction to Ichthyology* (Upper Saddle River, N.J.: Prentice Hall, 1996), p. 142.

17 *And since these signals must be visible, the fur withdrew. Our faces are bare so others can read them.*

The brain changed as well. For instance, the amygdala is a small cerebral organ that recognizes signals of anger and fear, and spurs production of hormones like adrenaline in response. In prosimians it abounds with connections to the olfactory sense. But in monkey, ape, and human brains, these links wane and far more arise to the visual cortex. As neuroscientist Leslie Brothers notes, "The transmitting equipment — expressive faces — and the receiving equipment — sensory brain wired to the amygdala — were evolving in step, prodded on by the demanding social milieu." *Friday's Footprint: How Society Shapes the Human Mind* (New York: Oxford University Press, 1997), p. 57.

19   *They may have evolved a two-legged stance to cross the ground from tree to tree more*
      quickly.

Bipeds move more efficiently than apes, though less so than four-legged
animals like dogs. Steven Stanley, *Children of the Ice Age* (New York: Harmony
Books, 1996). But see also Lynne A. Isbell and Truman P. Young, "The
Evolution of Bipedalism in Hominids and Reduced Group Size in
Chimpanzees: Alternative Responses to Decreasing Resource Availability,"
Journal of Human Evolution, vol. 30, pp. 389–397, 1996. The latter suggest that
the die-off of trees forced an evolutionary choice between smaller social
groups, the avenue of chimps, or larger groups, the route of prehumans.
Larger groups led to greater foraging range, and hence bipedalism.

19   *How did it stave off carnivores? It almost certainly used weapons. It could hurl rocks at*
      them, but more significantly, it made stone tools.

Pygmy chimps devise simple stone tools, and observers have seen even
capuchin monkeys make stone flakes and use them for cutting. Gregory
Charles Westergaard and Stephen J. Suomi, "A Simple Stone-Tool
Technology in Monkeys," *Journal of Human Evolution*, vol. 27, pp. 399–404,
1994. It's possible the australophithecines had stone tools as well, since as
Darwin notes, once prehumans stood on two feet, they could have better
"defended themselves with stones or clubs, or . . . attacked their prey." *The*
Descent of Man (Princeton, N.J.: Princeton University Press, 1871/1981),
p. 142. But these items become abundant and apparently crucial with *Homo*
habilis, whose name means "handy man."

20   *It differed strikingly from that of even the Neanderthals, our closest cousins.*

They were not, it seems, our direct ancestors. Geneticists have so far failed
to find any distinguishing DNA that we share with them.

20   *As Darwin suggested in* The Descent of Man, *our brains made the muzzle obsolete.*

"As [prehumans] gradually acquired the habit of using stones, clubs, or
other weapons, for fighting with their enemies, they would have used their
jaws and teeth less and less. In this case, the jaws, together with the teeth,
would have become reduced in size." *The Descent of Man*, p. 144.

23   *Yet the evolutionary stages are out in plain view.*

Richard Dawkins describes them in detail in *Climbing Mount Improbable*
(New York: W. W. Norton, 1996), pp. 138–197. The multifaceted insect
eye, essentially a motion detector, seems an evolutionary dead end.

23   *Monoculars like the bloody Cyclopeans of* The Odyssey *and the griffin-fighting*
      Arimaspi in Herodotus live only in myth.

And in genetic misfires. For instance, if pregnant ewes eat a certain skunk
cabbage found in the Rocky Mountains, they can give birth to cyclopeans
with one central eye, fused brain hemispheres, and no pituitary gland. These
damaged lambs die quickly. Scott F. Gilbert, *Developmental Biology*
(Sunderland, Mass.: Sinauer Associates, 1988), p. 188.

23   *Even primitive worms like the half-inch* Planaria *which dwells under rocks and in*
      streams, have paired eyes.

Amoebas, earthworms, and frogs are anti-Cyclopeans: They sense light
over their whole bodies.

24   *One scientist suggests this "attention structure" quickly clues us to hierarchy and fosters social coherence.*

M. Chance, "Social Structure of a Colony of *Macaca mulatta.*" *British Journal of Animal Behavior*, vol. 4, pp. 1–13, 1956. As Simon Baron-Cohen notes, "Gaze gives an instant snapshot of social status in a group." *Mindblindness* (Cambridge, Mass: MIT Press, 1995), p. 101.

27   *But eyes contact air directly, so to keep them wet, creatures developed eyelids with tear glands.*

Indeed, amphibian larvae lack eyelids and sprout them when they metamorphose into land creatures.

Nature abounds in exceptions to rules. Some fish have tearless eyelids. For instance, sharks possess a nictitating membrane which shields the eye in danger, as during attacks. At the same time, a few land animals lack blinking eyelids. Snakes and some reptiles have a transparent membrane called a "spectacle" over the eye. It is essentially fused lids, and bestows their famous stare. Twigs or claws can scratch the spectacle, so snakes shed it along with their skin when they molt.

32   *The nose has bemused anatomists.*

Tj. D. Bruintjes, A. F. van Olphen, and B. Hillen recently complained of this problem in "Review of the Functional Anatomy of the Cartilages and Muscles of the Nose," *Rhinoplasty*, vol. 34, June 1996, pp. 66–74. Earlier, Marc C. Dion, Bruce F. Jafek, and Charles E. Tobin noted, "The anatomy of the nose is poorly defined, even in standard anatomical texts." "The Anatomy of the Nose," *Archives of Otolaryngology*, vol. 104, March 1978, pp. 145–150.

34   *one new theory points to* Helicobacter pylori, *a bacterium that causes ulcers.*

In a study of 31 rosacea victims, 84 percent had *H. pylori* in their stomachs, compared to 50 percent of the general population. A. Rebora, F. Drago, and A. Parodi, "May *Helicobacter Pylori* Be Important for Dermatologists?" *Dermatology*, vol. 191, 1995, pp. 6–8. This theory explains why rosacea victims often also suffer from gastritis, why stomach ulcers and rosacea seem to rise and fall in prevalence together, and why metronidazole and clarithromycin improve both conditions. See also Lawrence Charles Parish and Joseph Witkowski, "Acne Rosacea and *Helicobacter Pylori* Betrothed," *International Journal of Dermatology*, vol. 34, no. 4, April 1995, pp. 236–237.

35   *the owl has a nasal tuft that covers 30 degrees of its visual field.*

T. G. R. Bower, *Development in Infancy* (San Francisco: W. H. Freeman, 1974). The owl is one of the few flat-faced animals aside from us, and uses its face as a scoop to enhance hearing. Bower's notion finds some support in Alistair P. Mapp and Hiroshi Ono, "The Rhino-Optical Phenomenon: Ocular Parallax and the Visible Field Beyond the Nose," *Vision Research*, vol. 16, no. 7, pp. 1163–1165, 1986.

36   *In 1960 biologist Alistair Hardy suggested an even more radical notion, one writer Elaine Morgan has since expanded on.*

Alistair Hardy, "Was Man More Aquatic in the Past?" *New Scientist*, vol. 9, pp. 642–645, 1960. Among Morgan's works, see, e.g., *The Scars of Evolution: What Our Bodies Tell Us about Human Origins* (London: Souvenir Press, 1990).

38   *the gateway for food, drink, and at times air.*

The starfish seems to reverse the flow. It pries a mussel apart, then thrusts its stomach out through its mouth to digest the creature. Food only enters its mouth when the stomach retreats. Christopher McGowan, *The Raptor and the Lamb* (New York: Henry Holt, 1997), p. 75.

40   *Most other animals use them mainly to seize prey.*

Snakes show the extreme here. Their incurved tines can snare and hold creatures, but do nothing more, so snakes swallow animals whole, in lurches.

40   *But mammal teeth can shear, crush, and grind food. They allow chewing, a notable power.*

Primates have a single, fused mandible, and it lets them use muscles on both sides of the face to crush food between teeth on a single side, thus increasing force. For more on the importance of chewing, see Milton Hildebrand, *Analysis of Vertebrate Structure* (New York: John Wiley & Sons, 1995), p. 115.

40   *a paleontologist can often identify a genus and even a species from a single tooth.*

Archeologists can also read diet in the patterns of wear on the teeth.

For instance, before the invention of pottery around 5300 B.C., teeth show large pits and much abrasion, from grinding dry, hard grains. Pottery enabled boiling and thus porridge, which left teeth much smoother. Porridge also let parents wean children earlier, before the emergence of their milk teeth, and hence boosted the birthrate and population density. "This is the major consequence of the so-called agricultural revolution." Theya Molleson, Karen Jones, and Stephen Jones, "Dietary Change and the Effects of Food Preparation on Microwear Patterns in the Late Neolithic of Abu Hureyra, Northern Syria," *Journal of Human Evolution*, vol. 24, pp. 455–468, 1993, p. 465.

41   *in Mesoamerica alone, dental anthropologists have identified 59 different types of tooth mutilation.*

For pictures of them, see George R. Milner and Clark Spencer Larsen's interesting "Teeth as Artifacts of Human Behavior," in Marc A. Kelley and Clark Spencer Larsen, eds., *Advances in Dental Anthropology* (New York: John Wiley & Sons, 1991), p. 359.

42   *Fish lack tongues.*

A few have tongue-like organs that don't move. Peter B. Moyle and Joseph J. Cech, Jr., *Fishes: An Introduction to Ichthyology* (Upper Saddle River, N.J.: Prentice Hall, 1996).

42   *We think of the tongue as a single muscle, but it is actually a bundle of them which can shorten, lengthen, and widen it.*

Four muscles lie on each side, arrayed in pairs: the superior longitudinal, the inferior longitudinal, the transverse, and the vertical. The longitudinal muscles pull the tip back, shortening, broadening, and thickening the tongue. The transverse muscles lengthen, narrow, and thicken it. The vertical muscles also lengthen the tongue, but broaden and flatten it, as when we stick it far out. G. J. Romanes, ed., *Cunningham's Textbook of Anatomy* (Oxford: Oxford University Press, 1981), p. 296.

43   *The nasolabial folds vanish in a paralyzed face, but not in a corpse.*

Lawrence R. Rubin, Yousri Mishriki, and Gene Lee, "Anatomy of the Nasolabial Fold: The Keystone of the Smiling Mechanism," *Plastic and*

Reconstructive Surgery, vol. 83, pp. 1–8, January 1989. Very little research so far has addressed these folds.

43   *The chin is unique to humans. Not even Neanderthals had one.*

They did possess its rudiments, a mini-chin. But archeologists consider a chin diagnostic of modern *Homo sapiens*, an interesting fact considering the near-ubiquity of chins on science-fiction species from different solar systems like the Klingons.

43   *It seems to aid mastication,*

Donald H. Enlow, *Facial Growth* (Philadelphia: W. B. Saunders, 1990), p. 298.

43   *its arrival coincides with no known shift in diet or subsistence pattern.*

Y. M. Lam, O. M. Pearson, and Cameron M. Smith, "Chin Morphology and Sexual Dimorphism in the Fossil Hominid Mandible Sample from Klasies River Mouth," *American Journal of Physical Anthropology*, vol. 100, pp. 545–557, 1996, p. 554.

44   *Since testosterone weakens the immune system, its excess is an "honest" signal of good health in men, advertising their ability to resist disease.*

For further references, see Steven W. Gangestad, Randy Thornhill, and Ronald Yeo, "Facial Attractiveness, Developmental Stability, and Fluctuating Asymmetry," *Ethology and Sociobiology*, vol. 15, pp. 73–85, 1994.

Whatever its uses for most people, the chin can be crucial for quadriplegics. Technology now allows their chins to control trackballs for typing and the joysticks that drive powered wheelchairs. See Reinhilde Jacobs, Elke Hendrickx, Isabel van Mele, Kirtire Edwards, Mieke Verheust, Arthur Spaepen, and Daniel van Steenberghe, "Control of a Trackball by the Chin for Communication Applications, with or without Neck Movements," *Archives of Oral Biology*, vol. 42, no. 3, pp. 213–218, 1997.

49   *Christopher Nyrop, author of* The Kiss and Its History

(London: Sands, 1901). Highly recommended.

51   *A kissing bandit of a different sort was twenty-two-year-old Tabetha Dougan.*

The court sentenced her to fifty years in prison. "Kissing Bandit Guilty of Picking up Men in Bars, Drugging Them," *San Jose Mercury News*, October 8, 1994, p. 22A.

60   *The third eye normally refers to an organ that perceives inner essence,*

It has not always been godly or metaphorical. Most early fish and reptiles possessed a physical third eye. It lay in the middle of the forehead or atop the head, and dwindled away in the Triassic. Today it exists only in lampreys and a few lizards, wholly covered by skin and serving mainly to detect the presence of light. But it has left a descendant: the pineal gland in the brain, which helps regulate circadian rhythms like sleep. Alfred Sherwood Romer and Thomas S. Parsons, *The Vertebrate Body* (New York: Harcourt Brace Jovanovich College Publishers, 1986), p. 518.

62   *Fish lack true ears. Rather, they have a lateral line system, which detects quivers in water.*

Some fish also hear with their swim bladders.

62   *(In a famously weird evolutionary move, the jawbones migrated upward to form these tiny structures.)*

Echolocating whales detect sound vibrations in their jawbone, which relays them to the ear. Likewise, some snakes touch their jaws to earth to hear. The jaw also served this secondary function in an ancestor of ours, and in a complex series of events, the task preempted it.

63   *For Muslim men, sporting a beard is* sunnat: *one earns credit for it, but suffers no penalty for abstaining.*

Myriad meanings attend beards in Islam, as everywhere else. During the *hajj,* the pilgrimage to Mecca, most men cease shaving to suggest their unworldly contemplation of Allah. In Turkey, by tradition only older men grew beards and they treated hairy-jawed young males as upstarts. The beard can convey political sentiments, and in recent years it has signaled one's nationalism and fundamentalism. The code can get fairly subtle In Turkey during the early seventies, conservatives turned the ends of their mustaches up and liberal radicals turned them down. Carol Delaney, "Untangling the Meanings of Hair in Turkish Society," *Anthropologica Quarterly.* vol. 67, no. 4, pp. 159–172, October 1994.

70   *It's known as pseudofolliculitis barbae (PFB), or "beard bumps," and it resembles acne. The tiny beard hairs curl back and reenter the skin.*

Blacks tend to have curved follicles, so the hair emerges from the skin and arcs back toward it. Close shaving cuts the hair at a slanted angle, leaving a pointed tip that can pierce the skin.

PFB can also arise within the follicle itself. Some shaving techniques — cutting against the grain, using a twin-blade razor, or pulling the skin taut — force the hair outward and yield a closer crop. But after the blade passes, the hair snaps back below skin level, where its honed tip can penetrate the follicle and grow directly into the flesh. Beatriz H. Coquilla and Charles W. Lewis, "Management of Pseudofolliculitis Barbae," *Military Medicine,* vol. 160, no. 5, pp. 263–269, 1995.

One fairly new treatment for PFB involves applying a lotion containing glycolic acid twice a day. So far, it is unclear why this chemical works. Nicholas V. Perricone, "Treatment of Pseudofolliculitis Barbae with Topical Glycolic Acid: A Report of Two Studies," *Cutis,* vol. 52, pp. 232–235, October 1993.

73   *Why did beards appear at all?*

Beards and mustaches are not solely human. They also grace certain primates. The douc (*Pygathrix nemaeus*), a remarkable monkey from Indochina, has a face haloed ear-to-ear in hairs like cat's vibrissae. The male bearded saki (*Chiropotes satanas*) boasts a great half-moon of whiskers beneath its face, and De Brazza's monkey (*Cercopithecus neglectus*) has a white chin cascade. The latter creature lives near others in its genus, so the beard identifies it for potential mates and slashes the incidence of hybrids, which are normally less fertile. J. R. Napier and P. H. Napier, *The Natural History of Primates* (Cambridge, Mass.: MIT Press, 1994), p. 140. Beards may perform this service for other monkeys as well.

The mustache is a rarer item. The Sumatran orangutan (*Pongo pygmaeus abelii*) sports a smear of hair on its upper lip, along with a splendid forked beard. The white-collared mangabey (*Cercocebus torquatus*) has a thin mustache, and a few other monkeys also possess scanty specimens. No primate has the flowing mustache of *Homo sapiens.*

73   *British archeologist A. M. W. Porter suggests that, since perspiration works mainly when it evaporates, beards may be sweat-catchers, useful to hunting men but not to women.*

"A sweat drop which falls to the ground is a sweat drop wasted." Porter, "Sweat and Thermoregulation in Hominids," *Journal of Human Evolution*, vol. 25, pp. 417–423, 1993, p. 419. Chest hair also stops sweat, he says, and a thick upper lip, flared nostrils, and a protruding chin slow its descent.

73   *Beards may also work by suggesting the chin-jut, a threat among both humans and chimps.*

Frank Muscarella and Michael R. Cunningham, "The Evolutionary Significance and Social Perception of Male Pattern Baldness and Facial Hair," *Ethology and Sociobiology*, vol. 17, pp. 99–117, 1996, p. 101. Of course, this factor could also have enhanced the appeal of large chins themselves.

Panoply

80   *genes don't fully mold the face, and we don't even know how much they shape it.*

"Despite considerable research, rather little is known about the precise relative importance of genetic and nongenetic factors in the outcome and the growth of the craniofacial skeleton." Tuula Laatikainen, Reijo Ranta, and Rolf Nordstrom, "Craniofacial Morphology in Twins with Cleft Lip and Palate," *The Cleft Palate-Craniofacial Journal*, vol. 33, no. 2, pp. 96–103, March 1996, p. 96.

80   *A flawed gene apparently enables the defect,*

Research on mice strongly implicates the gene that codes for transforming growth factor-$\beta 3$, a protein apparently vital for normal development of lungs and palate. Vesa Kaartinen, Jan Willem Voncken, Charles Shuler, David Warburton, Ding Bu, Nora Heisterkamp, and John Groffen, "Abnormal Lung Development and Cleft Palate in Mice Lacking TGF-$\beta 3$ Indicates Defects of Epithelial-Mesenchymal Interaction," *Nature Genetics*, vol. 11, pp. 415–421, December 1995. See also Diego F. Wyszynski, Terri H. Beaty, and Nancy E. Maestri, "Genetics of Nonsyndromic Oral Clefts Revisited," *The Cleft Palate-Craniofacial Journal*, vol. 33, no. 5, pp. 406–417, September 1996.

Intriguingly, among female identical twins, one will occasionally have a genetic disease normally found only in males, such as muscular dystrophy, hemophilia, color blindness, or fragile X syndrome. The cause is X inactivation, in which some paternal or maternal genes go inert soon after conception. Their shutdown can void the genetic shield female carriers usually possess, so they develop the ailment themselves, become "manifesting carriers." X inactivation is more common in female identical twins than in singletons, and some researchers speculate that it may be a cause of twinning itself. See N. Abbadi, C. Philippe, M. Chery, H. Gilgenkrantz, F. Tome, H. Collin, D. Theau, D. Recan, O. Broux, M. Fardeau, J.-C. Kaplan, and S. Gilgenkrantz, "Additional Case of Female Monozygotic Twins Discordant for the Clinical Manifestations of Duchenne Muscular Dystrophy Due to Opposite X-Chromosome Inactivation," *American Journal of Medical Genetics*, vol. 52, pp. 198–206, 1994, p. 198.

80   *women who take folic acid slash the risk of it in their babies.*

Gary M. Shaw, Edward J. Lammer, Cathy R. Wasserman, Cynthia D. O'Malley, and Marie M. Tolarova, "Risks of Orofacial Clefts in Children Born to Women Using Multivitamins Containing Folic Acid Periconceptionally," *Lancet*, vol. 346, pp. 393–396, 1995. The women in this study took folic acid in multivitamin supplements, from two months before conception to three months after. The incidence of cleft palate fell by 25 to 50 percent.

80   *when doctors detect twins before ten weeks, single births ensue in up to 78 percent of cases.*

I. Nakamura, M. Uno, Y. Io, I. Ikeshita, K. Nonaka, and T. Miura, "Seasonality in Early Loss of One Fetus Among Twin Pregnancies," *Acta Geneticae Medicae et Gemellologiae: Twin Research*, vol. 39, pp. 339–344, 1990.

82   *But we do not come into the world fully adept at face recognition. The skill matures.*

For this reason and others, such as the fact that neural networks can identify faces, some scientists believe the brain is not prewired for face recognition, but learns the skill through repeated exposure. For instance, see Isabel Gauthier and Michael J. Tarr, "Becoming a 'Greeble' Expert: Exploring Mechanisms for Face Recognition," *Vision Research*, vol. 37 no. 12, pp. 1673–1682, 1997. The debate here reflects the larger issue of how much the brain is a universal learning device and how much a modular one.

83   *But this research led to a device for recognizing faces.*

Much of the information in this section comes from my interview with Sandy Pentland. But see also Matthew Turk and Alex Pentland, "Eigenfaces for Recognition," *Journal of Cognitive Neuroscience*, vol. 3, no. 1, pp. 71–86, 1991.

85   *The Fifth Marquess of Salisbury was notoriously bad at recognizing faces,*

In Antonio R. Damasio, Hanna Damasio, and Gary W. Van Hoesen, "Prosopagnosia: Anatomic Basis and Behavioral Mechanisms," *Neurology*, vol. 32, pp. 331–341, April 1982, p. 331.

86   *Prosopagnosics often have trouble distinguishing members of a class, like different kinds of animals, flowers, and foods.*

In Oliver Sacks's *The Man Who Mistook His Wife for a Hat* the title character had worse problems. He confused basic categories themselves, and tried to don his wife's head like a cap. (New York: Summit Books, 1985.)

88   *We may store names in the left middle temporal lobe.*

Justine Sergent, Shinsuke Ohta, and Brennan MacDonald, "Functional Neuroanatomy of Face and Object Processing," *Brain*, vol. 115, pp. 15–36, 1992, p. 28.

89   *We may call up the dossier from the front of the temporal lobe.*

Sergent et al., above, pp. 30–31.

90   *Indeed, some scientists think it may be a clearinghouse for all social signals.*

Neuropsychiatrist Leslie Brothers, who does not go this far, says, "Our brains are evolutionarily prepared to generate certain responses to particular social situations. These responses are encoded in links between sensory representations of social events, and bodily changes — links that are found especially in the amygdala, where stimulation may produce feelings that are specifically appropriate for social situations." *Friday's Footprints: How Society Shapes the Human Mind* (New York: Oxford University Press, 1997), p. 52.

90   *In another using PET scans, the cingulate gyrus showed heightened activity when sub-jects were distinguishing happy, sad, and neutral faces.*

Mark S. George, Terence A. Ketter, Debra S. Gill, James V. Haxby, Leslie G. Ungerleider, Peter Herscovitch, and Robert M. Post, "Brain Regions Involved in Recognizing Facial Emotion or Identity: An Oxygen-15 PET Study," *Journal of Neuropsychiatry*, vol. 5, no. 4, pp. 384–394, Fall 1993.

96   *Over the last centuries an inadvertent historical experiment occurred on tiny St. Vincent in the Antilles.*

In Jonathan Kingdon's intriguing *Self-Made Man and His Undoing* (New York: John Wiley & Sons, 1993), p. 251.

96   *The best current evidence indicates* Homo sapiens *left Africa around 100,000 years ago*

A minority of scientists holds that our species developed throughout most of the Old World over the last two million years.

97   *In the United States each year, whites average 4.1 to 4.4 cases per 100,000 people, blacks 0.6 to 0.7.*

William H. Durham, *Coevolution: Genes, Culture, and Human Diversity* (Stanford, Calif: Stanford University Press, 1991), p. 503.

98   *Scientists predict cases of skin cancer will double in the United States and Europe by 2100.*

Harry Slaper, Guus J. M. Velders, John S. Daniel, Frank R. de Gruil, and Jan C. van der Leun, "Estimates of Ozone Depletion and Skin Cancer Incidence to Examine the Vienna Convention Achievements," *Nature*, vol. 384, pp. 256–258, November 21, 1996, p. 257. And these scientists deem their estimate conservative. Maritza Perez, a dermatology professor at Columbia University, offers a graver forecast. One of every 105 people born in 1990 will contract melanoma, she suggests, and one of 75 of those born in 2000 will. Perez, "Advances in Dermatologic Surgery," *Dermatologic Clinics*, vol. 15, no. 1, pp. 9–18, January 1997, p. 10.

A vaccine that prevented melanoma would, of course, vaporize these pro-jections, and scientists are now working on one. So far it does not forestall or cure melanoma, but does delay its progress. See Christine A. Kuhn and C. William Hanke, "Current Status of Melanoma Vaccines," *Dermatological Surgery*, vol. 23, pp. 649–655, 1997.

99   *variations in two kinds of melanin account for almost all mammalian browns, reds, golds, and blacks in the hair and fur.*

Arthur Hook, D. S. Wilkinson, F. J. G. Ebling, R. H. Champion, and J. L. Burton, eds., *Textbook of Dermatology*, 4th ed., vol. 3 (Oxford: Blackwell Scientific Publications, 1986), p. 2020.

100   *this notion dubiously postulates that Asians needed it because they lived near snowfields.*

Anthropologist A. T. Steegmann, Jr., measured facial temperatures in the lab and found that the prominent cheekbones of Asians get colder, not warmer, in a 32° F. breeze. And the short Asian nose did not reduce frostbite, since nose temperature bore no relation to bulk or salience (though it did tend to increase with head size). He concluded that the Asian face did not evolve to fight chill. Steegmann, "Human Adaptation to Cold," in Albert Damon, ed., *Physiological Anthropology* (New York: Oxford University Press, 1975), pp. 130–165.

100   *In very cold or dry regions, noses tend to be longer and narrower.*

A. Thomson and L. H. D. Buxton, "Man's Nasal Index in Relation to Certain Climatic Conditions," *Journal of the Royal Anthropological Institute* (now called *Man*), vol. 53, pp. 92–122, 1923. Of course, individual noses vary in every climate.

111   *The earliest known mirrors are obsidian disks from Çatal Hüyük in Turkey, dating back to 6500–5700 B.C.*

LeRoy McDermott, "Self-Representation in Upper Paleolithic Female Figurines," *Current Anthropology*, vol. 17, no. 2, pp 227–275, 1996, p. 228.

112   *lovers see their image in each other's pupils, enlarged by the convexity.*

Our word "pupil" comes from the Latin *pupilla*, "little doll," after the tiny person in the eye.

113   *In one literary convention, reflection requires a soul. Dracula has forfeited his and can't see himself in the mirror.*

On the other hand, warriors in the New Guinea highlands deem their spirits so potent that they can imprint their faces on the mirror. Timothy Troy, "Anthropology and Photography: Approaching a Native American Perspective," *Visual Anthropology*, vol. 5, pp. 43–61, 1992, p. 49.

113   *Some chimps and orangutans go further. They use mirrors to examine their teeth, rumps, and other body parts that usually lie out of sight. They make comic faces, place vegetables atop their heads, even adorn themselves, draping vines around their necks.*

Frans de Waal, *Peacemaking Among Primates* (Cambridge, Mass.: Harvard University Press, 1989), p. 85.

114   *Gallup concluded that chimps and orangutans were "self-aware."*

He also argued intriguingly that "the emergence of self-awareness is equivalent to the emergence of mind." Gordon G. Gallup, Jr., "Self-Awareness and the Emergence of Mind in Primates," *American Journal of Primatology*, vol. 2, pp. 237–248, 1982, p. 245.

118   *Totem poles could depict important deeds in a family's past,*

A true "totem" is a species or kind of object invested with awe, and the tribe never kills its individuals for food. In this sense, the Northwest totem poles don't involve totems, since images stood for distinct supernatural beings. The Indians felt free to dine on the animals.

119   *A more lurid case involves artist Oskar Kokoschka (1886–1980).*

Alfred Weidinger, *Kokoschka and Alma Mahler* (New York: Prestel, 1997).

121   *Salvador Dali's Mae West's Face Which Can be Used as a Surrealist Apartment (1933–1935) depicts West as a pure creature of film.*

To an extent, she was. She looks much more glamorous in her films than in photos from her youth, and the credit goes to makeup artist Dot Ponedel and especially cinematographer Karl Struss. Struss once stated that his job was to "minimize . . . physical imperfections. Makeup, diffusion, lighting, and carefully chosen angles are the chief tools." He typically illumined the narrower half of West's face, thus slimming it overall, and he pioneered the soft-focus lens, which hazed her in a romantic blur. Struss had been a noted still-portrait photographer, a fact that explains the static poses West sometimes strikes in her films. See Emily Wortis Leider, *Becoming Mae West* (New York: Farrar, Straus & Giroux, 1997), pp. 291–292.

127   *The face is the classic power icon.*

"Icon" comes from the Greek *eikon*, or "image," by way of the ikons of the Greek Orthodox Church, portraits which partook of divinity.

132   *Actors' fees did soar, but overall revenues outpaced them.*

Ironically, by the thirties Laemmle deemed movie stars overpaid and refused to match their salaries from other studios. Partly as a result, Universal floundered.

144   *Robert Mauro and Michael Kubovy have produced evidence that the brain stores faces as caricatures.*

Mauro and Kubovy, "Caricature and Face Recognition," *Memory and Cognition*, vol. 20, no. 4, pp. 433–440, 1992.

Semaphore

182   *They are almost certainly unlearned. Thalidomide babies born blind, deaf, and armless show them.*

Irenaus Eibl-Eibesfeldt, "The Expressive Behavior of the Deaf-and-Blind-Born," in M. von Cranach and I. Vine, eds., *Social Communication and Movement* (San Diego, Calif.: Academic Press, 1973), pp. 163–194. Despite such evidence, a few psychologists claim we learn all facial expressions. Alan Fridlund of U.C. Santa Barbara deems the thalidomide study inconclusive because caretakers might have rewarded children's smiles with kisses and hugs. In definitive tests, he says, blind youngsters would "have blind caretakers. The studies would necessarily control for caretakers' touching the children's faces, or hearing their children's vocalizations." He admits such experiments face grave barriers. See his rich, contrary, and interesting *Human Facial Expression* (San Diego, Calif.: Academic Press, 1994), p. 113.

Apart from universality, other issues quicken this field. How clear is the distinction between voluntary and involuntary expressions? Do expressions show emotions or predispositions to act? Is "emotion" even a useful concept, scientifically? These controversies show a discipline very much alive and advancing.

182   *An angry face is a warning.*

Hence we profit from spotting it early, and one experiment showed an irate face "pops out" of a crowd of neutral faces. Christine H. Hansen and Ranald D. Hansen, "Finding a Face in the Crowd: An Anger Superiority Effect," *Journal of Personality and Social Psychology*, vol. 54, pp. 917–924, 1988. So does a staring face, though apparently less so. Michael von Grünau and Christina Anston, "The Detection of Gaze Direction: A Stare in the Crowd Effect," *Perception*, vol. 24, pp. 1297–1313, 1995.

185   *In fact, the link between odious tastes and the disgusted face is anatomical: The same part of the brain reacts to both.*

It is the anterior insular cortex. The researchers who discovered this fact note that "appreciation of visual stimuli depicting others' disgust is closely linked to the perception of unpleasant tastes and smells." M. L. Phillips, A. W. Young, C. Senior, M. Brammer, C. Andrews, A. J. Calder, E .T. Bullmore, D. I. Perrett, D. Rowland, S. C. R. Williams, J. A. Gray, and A. S. David,

"A Specific Neural Substrate for Perceiving Facial Expressions of Disgust," *Nature*, vol. 389, no. 2, pp. 495–498, October 1997, p. 496.

186   *Indeed, once we learn disgust of certain foods, like shellfish, we rarely unlearn it.*

Paul Rozin and April E. Fallon, "A Perspective on Disgust," *Psychological Review*, vol. 94, pp. 23–41, 1987, p. 38.

188   *People who have lived with a suffering individual assess it better, while clinicians surrounded by pain discount it even more, an important fact they may not realize.*

Kenneth M. Prkachin and Kenneth D. Craig, "Expressing Pain: The Communication and Interpretation of Facial Pain Signals," *Journal of Nonverbal Behavior*, vol. 19, pp. 191–205, 1994, p. 201.

197   *One is Johnmarshall Reeve of the University of Wisconsin, who found the eyes crucial to the expression of interest. . . . In his experiment, people identified the face of interest from film clips, which are probably essential to this kind of research.*

As he notes, eye behavior in particular defies the still photo. Reeve, "The Face of Interest," *Motivation and Emotion*, vol. 17, no. 4, pp. 353–375, 1993, p. 355.

197   *How many kinds of messages can the eye convey? We don't know. Psychologist Simon Baron-Cohen suggests fifteen spectra,*

Mindblindness: An Essay on Autism and Theory of Mind (Cambridge, Mass.: MIT Press, 1995), p. 114.

200   *The versatile Niels Stensen (1638–1686) solved the problem in 1662. By dissecting corpses, he discovered the main lacrimal gland, just above the outside corner of the eye.*

Little known today, Stensen also proved that the heart and tongue were muscles, and discovered the duct of the parotid gland, the largest of the three salivary glands in the mouth. He was an anatomist, geologist, crystallographer, paleontologist, and zoologist, and eventually became Bishop of Hamburg. Raffaello Cioni, *Niels Stensen, Scientist-Bishop* (New York: P. J. Kennedy & Sons, 1962).

201   *Darwin observed that the English wept less than Mediterraneans.*

One recent survey found that Israelis cry less than Britons, possibly, the authors suggest, because of mandatory military service in Israel, which "encourages an active and resourceful approach to the solution of problems, and produces someone better able to cope with difficulties." D. G. Williams and Gabrielle H. Morris, "Crying, Weeping, or Tearfulness in British and Israeli Adults," *British Journal of Psychology*, vol. 87, pp. 479–505, 1996, p. 503.

201   *Women cry longer and more often than men.*

They also use tears more commonly as a coping device, to win sympathy and shame malefactors. See, e.g., Filip De Fruyt, "Gender and Individual Differences in Crying," *Personality and Individual Differences*, vol. 22, no. 6, pp. 937–940, 1997.

204   *Likewise at the end of a Rocky, when tension vanishes in a golden haze of glory, the body suddenly does not need stress chemicals, and we force them out in tears.*

Movie weepers feel much more stress at the final scene than nonweepers do. Susan M. Labott and Randall B. Martin, "Weeping: Evidence for a Cognitive Theory," *Motivation and Emotion*, vol. 12, no. 3, pp. 205–216, 1988.

205   *Though courtroom judges are equally likely to find smilers and nonsmilers guilty, they give smilers lighter penalties, a phenomenon called the "smile-leniency effect."*

Marianne LaFrance and Marvin A. Hecht, "Why Smiles Generate Leniency," *Personality and Social Psychology Bulletin*, vol. 21, pp 207–214, 1995.

209   *Even microbes have a sense of play.*
Hiram Stanley, "Remarks on Tickling and Laughter," *American Journal of Psychology*, vol. 9, pp. 235–240, 1898, p. 236.

216   *scientists have since confirmed that comedy kills pain. So does tragedy.*
Dolf Zillmann, Steve Rockwell, Karla Schweitzer, and S. Shyam Sundar, "Does Humor Facilitate Coping with Physical Discomfort?" *Motivation and Emotion*, vol. 17, pp. 1–21, 1993, p. 17.

218   *Glenn Weisfeld of Wayne State University has advanced a tentative theory*
Weisfeld, "The Adaptive Value of Humor and Laughter," *Ethology and Sociobiology*, vol. 14, pp. 141–169, 1993.

219   *in one surprising study, Robert Provine of the University of Maryland showed that over 80 percent of laughs in normal conversation had nothing to do with humor.*
Provine, "Laughter Punctuates Speech: Linguistic, Social, and Gender Contexts of Laughter," *Ethology*, vol. 95, pp. 291–298, 1993.

220   *smiles, disgust, and sadness are contagious too.*
See Lars–Olav Lundqvist and Ulf Dimberg, "Facial Expressions Are Contagious," *Journal of Psychophysiology*, vol. 9, no. 3, pp. 203–211, 1995.

223   *Annoyingly, blushes love an audience.*
One study showed we blush more deeply as audience size grows from one person to four, though, intriguingly, not from zero to one. Don Shearn, Erik Bergman, Katherine Hill, Andy Abel, and Lael Hinds, "Blushing as a Function of Audience Size," *Psychophysiology*, vol. 29, no. 4, pp. 431–436, 1992.

224   *women overall are more emotionally responsive than men, especially in the face.*
In Scott R. Vrana, "The Psychophysiology of Disgust: Differentiating Negative Emotional Contexts with Facial EMG," *Psychophysiology*, vol. 30, pp. 279–286, 1993, p. 280.

224   *Investigators have suggested many reasons for the peak at adolescence, such as "self-con-sciousness and embarrassment following rapid bodily changes; hormonal changes; unsettled identity, particularly sexual identity; new kinds of social encounters which provide the setting for interpersonal judgments."*
Stephanie A. Shields, Mary E. Mallory, and Angela Simon, "The Experience and Symptoms of Blushing as a Function of Age and Reported Frequency of Blushing," *Journal of Nonverbal Behavior*, vol. 14, pp. 171–187, 1990, p. 183.

225   *Darwin wondered why blushes don't spread over the whole body, and suggested it's because the face commands our attention. Blushing is a signal, and the face is the body's best relay point.*
Some scientists have disagreed. They note that the face has more capillary loops than elsewhere, hence greater blood flow. Thus the face also reddens from noncommunicative causes—exercise, drinking hot liquids, the hot flashes of menopause, severe hypertension, illness—and the appearance of blushes on the face may be happenstance. On the other hand, studies also show that a blush raises the temperature of the face, but not of the rest of the body. The physiology of blushing remains poorly understood. See W. D. Cutlip II and M. R. Leary, "Anatomic and Physiological Bases of Social Blushing: Speculations from Neurology and Psychology," *Behavioural Neurology*, vol. 6, pp. 181–185, 1993.

227   *In one recent study 30 percent of blacks said "no one ever notices [my blushing]' and just 22 percent said others see a blush through change in skin color.*

Angela Simon and Stephanie A. Shields, "Does Complexion Color Affect the Experience of Blushing?" *Journal of Social Behavior and Personality*, vol. 11, pp. 177–188, 1996, p. 185.

229   *Monkeys reared in total isolation try to conciliate a staring, full-face picture more often than a profile.*

Morton J. Mendelson, Marshall M. Haith, and Patricia J. Goldman-Rakic, "Face Scanning and Responsiveness to Social Cues in Infant Rhesus Monkeys," *Developmental Psychology*, vol. 18, pp. 222–228, 1982.

229   *In one remarkable study, baby chicks shunned a darkened circle more than a rectangle, and a pair of circles more than one. And if the twin discs seemed to follow them, the chicks stayed even further away.*

M. Scaife, "The Response to Eye-Like Shapes by Birds," Parts I and II, *Animal Behavior*, vol. 24, pp. 195–206, 1976.

229   *But if another person enters, the very nature of the park seems to change. A second force field appears, and we go from watcher to watched, emperor to environment, willy-nilly.*

Psychoanalytic theorist Jacques Lacan (1901–1981) went further and argued that *physical objects stare back at us.* The idea came to him after he'd sailed with fishermen on the English Channel. One pointed to a floating sardine can and cried, "See that can there? Well, it doesn't see you!" Lacan pondered this statement and eventually came to disagree. Cans and rocks and tree stumps do look back, he said. Unfortunately, this arresting claim hides a fairly ordinary idea, since Lacan meant only that we perceive every object simultaneously with its web of associations. Thus we can't separate our sensation of a bobbing tin from our cultural preconceptions of "sardine can." The world "looks back at" us with all its acquired meanings.

232   *Indeed the symbol of Hussein's secret police is an eye on a map of Iraq.*

The Pinkerton Detective Agency adopted a staring orb as its emblem over a century ago, in 1850, along with the motto: "The Eye That Never Sleeps." Hence founder Allan Pinkerton (1819–1894) earned the nickname "The Eye," and this usage bred the term "private eye." Frank Morn, *"The Eye That Never Sleeps": A History of the Pinkerton National Detective Agency* (Bloomington, Ind.: Indiana University Press, 1982).

246   *police detectives performed no better than ordinary citizens, but to the researchers' surprise, they were confident they had.*

Aldert Vrij, "The Impact of Information and Setting on Detection of Deception by Police Detectives," *Journal of Nonverbal Behavior*, vol. 13, pp. 117–136, 1994.

247   *Only the Secret Service agents scored significantly above chance,*

Paul Ekman and Maureen O'Sullivan, "Who Can Catch a Liar?" *American Psychologist*, vol. 46, no. 9, pp. 913–920, September 1991, p. 917.

248   *Paul Ekman, in his book* Telling Lies,

(New York: W. W. Norton, 1992.) The best popular book on the subject.

251   *It meant the apparently unflappable Poindexter may have been hiding a key fact about this meeting, but interrogators missed it and didn't follow up.*

In 1990 a federal court convicted Poindexter of lying to Congress, as well as conspiracy and obstruction of justice.

260   *Acting changes from stage to screen,*

Likewise, actors onstage must project their voices to the back row, but in film they mute down. Actors who project on film will seem to bellow.

Siren

278   *In Shakespeare and Fletcher's* The Two Noble Kinsmen, *Palamon and Arcite spy a woman from a second-story window,*

In Chaucer's "The Knight's Tale" these two quarrel similarly, but less comically.

282   *So far, however, most studies have not broken down their results by these types.*

Though see psychologist Leslie A. Abramowitz's extended discussion of the babyface in *Reading Faces: Window to the Soul?* (Boulder, Colo.: Westview Press, 1997).

287   *It may affect as much as 2 percent of the population, or 5 million Americans.*

Katharine A. Phillips, *The Broken Mirror: Understanding and Treating Body Dysmorphic Disorder* (New York: Oxford University Press, 1996), p. 6. An authoritative, nontechnical account.

289   *even when they find a doctor who will perform it, they rarely like the outcome.*

BDD patients are a notorious vexation to doctors, often clogging the waiting rooms, phoning ceaselessly, filing lawsuits. Their behavior can get worse. "In recent times in the United Kingdom, one dermatologist and two plastic surgeons have been murdered, and practitioners working in this field should know there is a small but definite risk of assault when managing these patients." John A. Cotterill, "Body Dysmorphic Disorder," *Dermatologic Clinics,* vol. 14, no. 3, July 1996, pp. 457–463, p. 459.

304   *The Jivaro Indians of Ecuador long specialized in head-shrinking,*

Michael Harner, *The Jivaro: People of the Sacred Waterfalls* (Garden City, N.Y.: Doubleday/Natural History Press, 1972).

306   *Laser-wielding surgeons can now extirpate*

Laser surgery is advancing quickly and today's state-of-the-art is tomorrow's junkyard. But lasers have proved very effective at removing items once thought permanent.

The pioneer device was the carbon dioxide laser. This scalpel of light cut out warts and other skin growths, but its main advantage lay in the way it cauterized nearby tissue, reducing bleeding. Like other kinds of surgery, it left detectable scars and required a long healing period, and newer kinds of laser have replaced it.

Today, the reigning tool is the Q-switched laser, which pelts the skin with bursts of light measured in nanoseconds. ("Q-switching" is the technique that creates these pulses.) The ultrabrief exposure keeps heat from flowing into adjacent tissue and burning it. The hue of the laser beam limits damage too. Colored ink selectively absorbs light of a particular wavelength, so carbon-based tattoos, for instance, soak up the red beam of the Q-switched ruby laser. Scientists are unsure exactly why the laser works, but they believe it breaks the ink into small pieces, which scavenger cells mop up. See Suzanne Linsmeier Kilmer, "Laser Treatment of Tattoos," *Dermatologic Clinics*, vol. 15, no. 3, July 1997, pp. 409–417.

316   *we have far more sebaceous glands than other animals.*

For instance, primates have some, but far fewer than we do. J. R. Napier and P. H. Napier, *The Natural History of Primates* (Cambridge, Mass.: MIT Press, 1994), p. 31. They also appear in such mammals as dogs, cats, goats, and hamsters.

317   *one scholar calls it "the most significant therapeutic advance in acne therapy."*

James J. Leyden, "Oral Isotretinoin," *Dermatology*, vol. 195, suppl. 1, pp. 29–33, 1997, p. 29. The drug attacks acne on several fronts at once. Physicians once prescribed it mainly for severe cases, but they now offer it much earlier and for less serious conditions. Isotretinoin has predictable side effects, such as dry lips and eyes, which moisturizers can easily counter. Severe side effects are rare, but women of childbearing age must use contraceptives. See W. N. Meigel, "How Safe Is Oral Isotretinoin?" *Dermatology*, vol. 195, suppl. 1, pp. 22–28, 1997.

There are other therapies. Many scientists are excited about the new topical retinoids adapalene and tazarotene. Surprisingly, even visible light helps to some extent. V. Sigurdsson, A. C. Knulst, and H. van Weelden, "Phototherapy of Acne Vulgaris with Visible Light," *Dermatology*, vol. 194, suppl. 1, pp. 256–260, 1997.

323   *Tagliacozzi himself was a popular surgeon, and his method for rebuilding noses lost in a duel might have helped the Phantom.*

Historians once thought Tagliacozzi invented this technique, but the laurels go to predecessors such as Gustavo Branca (15th cen.) and Leonardo Fioravanti (1518–1588). Branca himself may have learned it in Persia. Nevertheless, Tagliacozzi made important contributions to it. See Paolo Santoni-Rugiu and Riccardo Mazzola, "The Italian Contribution to Facial Plastic Surgery: A Historical Reappraisal," *Facial Plastic Surgery*, vol. 12, no. 4, pp. 315–320, October 1996.

324   *Ronald Matsunaga, a Beverly Hills plastic surgeon who specializes in these operations, denies the charge.*

Laura Accinelli, "Eye of the Beholder," *Los Angeles Times*, January 23, 1996, pp. E1, E4.

333   *Indeed,* Homo sapiens *lived beside the Neanderthals for thousands of years, and we differed from them mainly in the face.*

In fact, Neanderthal faces eventually came to look more Neanderthal, as if mate recognition mattered more.

336   *women's fertility drops faster than men's over a lifetime.*

Doug Jones, "Sexual Selection, Physical Attractiveness, and Facial Neoteny," *Current Anthropology*, vol. 36, no. 5, pp. 723–743, December 1995, p. 726.

340   *Researchers mated the most attractive face with one of the other 29.*

They picked this second face by lot, weighted by looks. The process resembled a spinning wheel, in which each of the 29 had an area on the rim proportional to its beauty and researchers chose the breeder by, in effect, twirling the wheel. Victor Johnston and Melissa Franklin, "Is Beauty in the Eye of the Beholder?" *Ethology and Sociobiology*, vol. 14, pp. 183–199, 1993.

341   *If you are a female and all other females favor an average face in a mate, then you should too,*

Female Trinidadian guppies mate with certain males solely because they've seen other females favor them, and female sage grouse, river bullheads, and fallow deer may act similarly. See Stephen Pruett-Jones, "Independent Versus Nonindependent Mate Choice: Do Females Copy Each Other?" *The American Naturalist*, vol. 140, no. 6, December 1992, pp. 1000–1009. Pruett-Jones offers a different explanation, a game theory model in which imitation reduces the cost to females of seeking a fit mate. Either way, mate-preference copying plainly occurs in humans of both sexes, and may partly explain the allure of fame and prestige.

341   *Beauty is not a single quality. It involves a mix of signals, especially of averageness, sexual maturity, youth, and health.*

What about symmetry? It is crucial to mating success in certain animals, especially flying ones. Indeed, female scorpion flies may completely spurn a male if his forewings differ in length by a single millimeter. Hence some scientists suggested that symmetry makes people's faces more attractive as well, and one experiment found that it did. Karl Grammer and Randy Thornhill, "Human (*Homo sapiens*) Facial Attractiveness and Sexual Selection: The Role of Symmetry and Averageness," *Journal of Comparative Psychology*, vol. 108, pp. 233–242, 1994.

Unfortunately, follow-up studies failed to replicate this finding. In one, infants from four to fifteen months old paid more attention to attractive faces, but not to symmetrical ones. Curtis A. Samuels, George Butterworth, Tony Roberts, Lida Graupner, and Graham Hole, "Facial Aesthetics: Babies Prefer Attractiveness to Symmetry," *Perception*, vol. 23, pp. 823–831, 1994. In another, researchers pored over photos of famous beauties and found their faces often quite asymmetrical (in Samuels et al., above, p. 824). Psychologist Rotem Kowner discovered that people actually deemed mild facial asymmetry more attractive, especially in the young. Symmetry boosted appeal only among the elderly, perhaps, he suggested, because over time hardships like poverty can worsen the imbalance. Rotem Kowner, "Facial Asymmetry and Attractiveness Judgments in Developmental Perspective," *Journal of Experimental Psychology*, vol. 22, pp. 662–675, 1996.

Acknowledgments

Everything is in the face. It is infinite and a book is not, and this one must end. But not without mentioning the many people whose knowledge, talent, and grace contributed to it.

Especially helpful was Terry Brazell, an astute dermatologist who spent hours guiding me through her beloved specialty. She was endlessly patient and cheerful, and I owe her much.

Many other scholars and professionals freely gave me their time and I am grateful to them. They include: Paul Ekman, of U.C. San Francisco; Vicki Bruce, of the University of Stirling, Scotland; Sandy Pentland, of the MIT Media Lab; Dominic Massaro, of U.C. Santa Cruz; R. Glenn Northcutt, of the Scripps Institute of Oceanography, La Jolla; Barry Weissman, of the Jules Stein Eye Institute, UCLA; actor-director Barbara Tarbuck; and sculptor Christalene Loren.

In addition, I benefited from discussions with Dan Friedman of U.C. Santa Cruz, Bart Kosko of U.S.C., and David Leon of the California State University at Sacramento. John Adam, Paul Freiberger, Howard Gold, Janet Howard, Ken Johnson, Don

Kreuzberger, and Rob Polevoi aided this project in many ways. I also profited from the Gruter Institute Seminar on Law, Biology, and Human Behavior at Squaw Valley in June 1996 where I enjoyed warm hospitality from Dr. Margaret Gruter and the staff. Dan Friedman, David Leon, Andrea Polevoi, Rob Polevoi, and Barbara Tarbuck read portions of the book and offered valuable comments.

I am grateful to Mike, Sue, and Bill Elwell, whose outstanding company helped revive me after one especially long and savage bout of writing.

I am blessed with splendid agents in Katinka Matson and John Brockman, who encouraged this project from first gleam and guided it wisely to fulfillment. Sharon Constant, of Visible Ink in Oakland, drew the illustrations, and she's a true pro. And at Little, Brown, I'm grateful to Mike Liss, copyeditor Betty Power, and especially my keen-eyed editor, Rick Kot. He was both enthusiastic and shrewd about the manuscript, and his intelligence has sharpened it throughout.

To you all, I turn my own face and bow.

Index